Understanding Organizational Behaviour and Human Resources Management

NIPA® GENX ELECTRONIC RESOURCES & SOLUTIONS P. LTD.
New Delhi-110 034

About the Editors

K. Ekta is a nominated 'Author of the Year Award' by StoryMirror for her literary work *'Sang Jeene Ke Liye'*. She is an MBA in Human Resource Management from Pondicherry University (Central University, Govt. of India) and has published and presented her academic work in reputed journals and conferences. She is currently a Freelance Consultant and Patron of Development Connects, a consulting firm at Ranchi, Jharkhand. She has worked on public communication, counselling, training and development.

Sanjeet Kumar Sameer is currently Associate Professor at School of Agribusiness and Rural Management, Dr. Rajendra Prasad Central Agricultural University, Pusa, Bihar (An Institution of National Importance). He completed his PhD in Organizational Behaviour and Human Resource Management from Indian Institute of Management Lucknow and MBA from Institute of Rural Management Anand after his B.Tech from NDRI, Karnal. He is a certified Business Administration / Business Economics Professional from BWI, Basel, Switzerland. He has more than 17 years of experience in corporate and academia and across multiple business verticals and projects including at *Fortune 500* Company.

About other Contributors

Abhishek Tiwary is the Senior Vice President and Global Head-HR (BPS) for Tech Mahindra. He joined Tech Mahindra in April 2023. Prior to Tech Mahindra, he served at KPMG for 15 years – in his last appointment at KPMG, he served as the Global Head-HR for Advisory business based out of their New York office. As a Global HR executive with a career spanning over 20 years, Abhishek has run HR functions for global enterprises through their growth journeys. Having worked in a variety of sectors including Technology, Consulting, Professional Services, Media and the Government, Abhishek has played a key role in Culture building, Organizational Development, DEI frameworks and big data analytics to drive people decisions. Throughout his career, he has endeavoured to lead business leaders into adopting a 'people-first' mindset for sustainable business growth. He believes that this has helped create thriving organizational ecosystems that empower people to live and work with pride and passion. He is passionate about Diversity, Equity and Inclusion and has sought to create inclusive workplaces. Abhishek started his career with the Indian Navy and is a proud alumnus of the National Defence Academy, Khadakwasla. He holds a Masters degree in Personnel Management from SIBM, Pune. With a keen interest in the interplay between behavioral science, evolving workplaces and technology, he is currently pursuing PhD in organizational behavior from IIM Lucknow. His research is focused on causal relationships between automation technologies, role of the workplace and outcomes for employees.

Pinaki Chakladar is an Organizational Learning and Change Management Consultant, who is also IBM-certified in 'Digital Change' and 'Organizational Learning.' He is also a specialist in the ISD (Instructional Systems Design) methodology and a wide variety of Technology-based Learning systems and methods. In addition, he holds functional domain experience in Performance, Talent and Knowledge Management. Pinaki holds more than 24 years of experience spanning multiple domains, such as Organizational Learning, Leadership development and Change Management. He also holds expertise in other allied areas, such as Performance and Talent Management and various facets of Knowledge Management (including KM strategy, and Social Networking Analysis (SNA). Additionally, he is PMP-certified and holds project/program management skills. Pinaki is an alumnus of IIM Lucknow. Currently he is Director at IBM India Pvt. Ltd.

Understanding Organizational Behaviour and Human Resources Management

K. Ekta
Sanjeet Kumar Sameer

NIPA® GENX ELECTRONIC RESOURCES & SOLUTIONS P. LTD.
New Delhi-110 034

The cases used in the book have been developed in the form of corporate anecdotes with a view to help students and early stage practitioners in applying the concepts of Organizational Behaviour and Human Resource Management in a particular context without drawing similarity with other organization. All names in the cases and anecdotes have been changed to protect the identity of actual people, organisation and place. Any resemblance is purely coincidental. The views expressed by the authors and contributors are personal and not of their current or previous employers or organizations and that of the publishers or their staff.

NIPA® GENX ELECTRONIC RESOURCES & SOLUTIONS P. LTD.

101,103, Vikas Surya Plaza, CU Block
L.S.C.Market, Pitam Pura, New Delhi-110 034
Ph : +91 11 27341616, 27341717, 27341718
E-mail:newindiapublishingagency@gmail.com
www: www.nipabooks.com

For customer assistance, please contact
Phone: + 91-11-27 34 17 17
Fax: + 91-11-27 34 16 16
E-Mail: feedbacks@nipabooks.com

Print ISBN: 978-93-58878-17-2

ebook ISBN: 978-93-58879-22-3

NIPA also publishes books in a variety of electronic formats. Some content that appears in print may not be available in electronic books, and vice versa.

Composed and Designed by NIPA.

Preface

Understanding organizational and human behaviour is a complex process. Similarly, managing human resources effectively is increasingly becoming a challenge for an organization. Having an in-depth knowledge of related concepts of organizational behaviour and human resource management is very much desirable for any manager so that these can be appropriately applied at their workplace.

Learning methods based on examples, stories, anecdotes and cases have been found to be very effective. This book has been written primarily with an intention to develop required understanding of the concepts of organizational behaviour and human resource management among students and early stage practitioners and demonstrate how these can be applied in different scenario by giving suitable examples and cases. Anecdotes from multiple sectors such as agriculture, IT, power, energy, consulting etc. have been used in this book. Only significant aspects of these concepts have been discussed in the book leaving aside detailed historical and research background of the same. A chapter on 'Responsible Organization' has been purposefully added in the book to sensitize readers about such requirements and highlighting a need for proactive preparedness for meeting future challenges.

Practice questions have been appended to each of these cases and anecdotes asking the readers to list down the concepts of OB and HRM that are relevant. They are expected to apply these concepts to analyse various situations and events of the case and anecdotes, explain how these are interrelated in the context of the case and propose possible solution(s) to the problem by applying these concepts.

This book is primarily meant for undergraduate/ post-graduate students and entry level practitioners but may not be very relevant for researchers including doctoral scholars and experienced human resource professionals.

Contents

Section I
Important Topics of Organizational Behaviour and Human Resource Management

1

Summary of Key Topics of Organizational Behaviour and Human Resource Management

K. Ekta

Freelance Consultant and Patron of Development Connects
A Consulting firm at Ranchi, Jharkhand

In simple words, Organisational Behaviour is the study of Individuals and Group behaviour within an organisation and indicates how these interactions affect an organization's performance toward its goals. It also examines the impact of various factors on behaviour within an organization and help in understanding and predicting organisational life.

Organizational Behaviour is a 'Specialised Stream' drawing inputs from multiple academic domains such as psychology, sociology, political science, anthropology, science, economics, and technology to name a few.

This chapter lists down various key words that are relevant from the perspective of Organizational Behaviour. The purpose of this book is not to describe each terms of OB as there are many books in the market that cover these aspects however it is always very relevant for students and practitioners to just get a glimpses of these key words and based on requirement they can explore further.

Concepts Associated with Understanding Organization

1. Why organization exists
2. Emerging challenges and expectations for organizations including diversity, inclusivity, innovation, adaptation and change management
3. Talent management
4. Internationalisation
5. Technology centric work design
6. Sustainability
7. Privacy
8. Work-life integration and balance

9. Organizational Behaviour (OB)
10. Evolution of OB
11. OB Models
12. Trends in OB
13. Associated streams of OB
14. Organization structure
15. Elements of organizational structure-work specialisation, departmentalisation, chain of command, span of control, centralisation, formalisation
16. Organizational design types-simple organization, bureaucracy, matrix structure, virtual organization, boundaryless organization
17. Organizational restructuring- lean organization, rightsizing
18. Models of organizational design
19. Organizational strategy, size, technology and environment
20. Organizational design and employee performance
21. Organizational culture- definition and perspectives
22. Organizational value
23. Uniform, non-uniform culture
24. Strong, weak culture
25. Subculture, Culture and formalization
26. Positive & Negative functions of culture
27. Organizational climate and role of culture
28. Organization and institutionalization
29. Culture creation and sustenance- Socialisation model
30. Transmitting culture to employees-stories, rituals, symbols, language, artefacts
31. Ethical organizational culture
32. Positive organizational culture
33. Spiritual organization
34. National culture- model
35. Organizational change and factors, planned and unplanned/ radical
36. Resistance to change
37. Theory of organizational change management: Lewin's 3- step model, Kotter's 8-step plan

38. Action research
39. Organizational development and techniques
40. Creating a culture for change
41. Creating a culture of innovation
42. Creating a learning organization
43. Work Stress, sources and consequences
44. Model of Stress
45. Stress and job performance
46. Stress management techniques
47. Future of organization

Concepts Associated with Understanding Individuals and Groups

1. Defining a manager
2. Functions of a manager
3. Managerial roles
4. Managerial skills and functions
5. Interpersonal skills
6. Successful and effective managers
7. Attitudes of a Manager- Types
8. Job attitudes
9. GLOBE
10. Affect
11. Emotions
12. Moods
13. Emotional labour
14. Cognitive dissonance
15. Emotional dissonance
16. Affective events theory
17. Emotional intelligence and models
18. Deviant workplace behaviours
19. Personality and determinants
20. The Big-Five Personality

21. MBTI
22. The Dark Triad
23. Proactive Personality
24. Values-Instrumental and Terminal
25. National Culture and Values
26. Perception
27. Attribution theory
28. Biases
29. Positive thinking
30. Mindfulness
31. Creativity
32. Positive enforcement
33. Decision making models: rational, bounded rationality and intuition
34. Ethical Decision criteria
35. Model of creativity
36. Motivation and theories of motivation
37. Self-efficacy
38. Organizational justice
39. Integration of theories of motivation
40. Motivating through job designing
41. Job characteristics model
42. Job redesigning and methods
43. Emerging trends in job redesigning- FTE, Job sharing, flexitime, telecommuting, Job crafting
44. Motivating through pay plans
45. Group and theory of group formation
46. Features/ Characteristics of roles of group
47. Decision making by group
48. Transactional analysis
49. Concepts related to groups- groupthink, group shift
50. Work groups and teams

51. Effective teams
52. Organizational demography
53. Communication & manager
54. Process and types of communication
55. Communication channel
56. Effective communication
57. Leadership
58. Leadership theories
59. Trait theories
60. Behavioural theories
61. Contingency theories: Fiedler model, situation leadership, path-goal theory, leadership-participation model
62. Leader Member Exchange (LMX) theory
63. Other types of leaders: charismatic leader, transformational leader, transactional leader, authentic leader, servant leadership
64. Mentoring, task and relationship oriented leadership, online leadership
65. Selecting leaders
66. Power and its basis
67. Power tactics
68. Political skills and organizational politics
69. Political behaviour- factors and consequences
70. Defensive behaviour and impression management
71. Political action & ethics
72. Conflict and perspectives of conflict
73. Conflict process
74. Conflict handling techniques
75. Conflict and performance
76. Conflict management styles
77. Negotiation and bargaining
78. Negotiation process
79. Effective negotiation
80. Third party negotiation

Human resource management describes the approaches involved in bringing people and organizations together so that both can achieve their goals. Like OB, this is also a specialized stream and primarily looks at human capital related issues in more detail in order to fulfill the goals of the organization.

In addition to the above topics, this chapter lists down various key words that are relevant from the perspective of Human Resource Management (HRM). The purpose of this book is not to describe each terms of HRM as there are many books in the market that cover these aspects however it is always very relevant for students and practitioners to just get a glimpses of these key words and based on requirement they can explore further.

1. Understanding basic management process
2. Definition of HRM
3. Importance of HRM
4. Challenges and trends shaping HRM
5. Line and staff function
6. Authority
7. Human capital
8. Personnel management Vs HRM
9. Strategic HRM
10. Sustainable HRM
11. Ethics in HRM
12. Digital HR
13. Employment laws in India
14. Prevailing labour laws in India
15. Labour Codes
16. Constitutional basis of labour laws
17. Seventh schedule entry related to labour laws/ labour
18. Labour laws enacted by central government and implemented by central government
19. Labour laws enacted by central government and implemented by both central and state governments
20. Labour laws enacted by central government and implemented by state government
21. Labour laws enacted by state government and implemented by state government

22. Articles affecting government employment
23. Features of labour legislation in India
24. Laws related to Industrial relationships
25. Laws related to wages
26. Laws related to working hours, conditions of services and employment
27. Laws related to equality and empowerment of women
28. Laws related to social security
29. Labour laws applicable to IT sector
30. Provisions/ laws related to sexual harassment
31. Administration of labour laws in India and related agencies
32. Managing diversity at workplace
33. Strategic HRM process
34. Types of strategy and application in HRM
35. Strategic HRM tools- HR / HRD Scorecard, HR Metrics, Benchmarking, HR Excellence model, Digital dashboard, HR Audits
36. High Performance Work Systems
37. Employee engagement
38. Talent Management Process
39. Job analysis, methods and application
40. Job description, job specification and job classification
41. Organogram / Organization chart
42. Workflow analysis
43. Job redesigning
44. Quantitative job analysis techniques: Position Analysis Questionnaire (PAQ), DOL procedure,
45. Job identification
46. HR Manager competency model
47. Competency statements
48. HR Planning and methods
49. Personnel / position replacement charts
50. Markov analysis: forecasting positions for internal employee
51. Forecasting supply of outside candidates

52. Succession Planning
53. Recruitment and application
54. Recruiting yield pyramid
55. Job posting- internal recruitment
56. Modes of recruitment including non-traditional methods
57. Outsourcing and off shoring of jobs
58. Employees selection methods/ techniques
59. Expectancy chart
60. Tests for cognitive abilities
61. Tests for personality and interests
62. Achievement tests
63. Situation judgement tests
64. Assessment centres
65. Job training and evaluation approach
66. Background check of candidates and methods
67. Interviews
68. Improving effectiveness of interview, how to conduct an effective interview
69. Competency based interview
70. Job offer
71. On-boarding and orientation
72. Designing training- Task Analysis, Training Needs Analysis, Competency oriented training
73. Conducting training programmes
74. Selecting employees for expensive training/development programmes: 9-Box Grid
75. Leadership Development and Life cycle learning
76. Managing organizational change and role of training/organizational development initiatives
77. Evaluation of training methods
78. Performance management, goals and methods
79. Techniques of performance appraisal

80. Issues in performance appraisal
81. Counselling and Feedback
82. Career management
83. Integrated talent management system
84. Role of coaching and mentoring in career management
85. Employee turnover management
86. Employee life-cycle career management: promotion, transfer, retirements, dismissals / terminations, termination and exit interviews, layoffs, retrenchment, closure, downsizing
87. Career choices
88. Job searching
89. Managing interviews
90. Compensation management and factors
91. Job evaluation and methods
92. Job classification
93. Designing pay plans-steps
94. Compensation for executives
95. Compensation for professionals
96. Trends in compensation management
97. Total rewards
98. Motivating employees through pay plans- Vroom's expectancy theory, Herzberg two factor theory, Skinner's behaviour reinforcement theory
99. Employee centric incentives and bonuses
100. Job design and employee motivation
101. Team centric incentives
102. Employee benefits including retirement benefits
103. Employee relations- definition and programs
104. Suggestion teams, quality circle, self-managed team
105. Ethical organization
106. Employee discipline management
107. Employee engagement surveys, job satisfaction surveys
108. Trade unions

109. Collective bargaining
110. Principles of bargaining
111. Dealing with impasses, mediation and strikes
112. Employee disputes and grievances
113. Health, Safety and Environment and HRM
114. Workplace health issues and remedies
115. Hazard/ Accidents prevention, Risk management
116. International HRM
117. Staffing systems in IHRM
118. Training and maintaining employees abroad
119. Pay parity : Parent company and foreign subsidiary
120. Repatriation
121. HRM for small businesses
122. Professional employer organizations, human resource outsourcing organization, staff leasing firms
123. Human Resource Information System
124. Future of HRM

Further Reading

https://open.umn.edu/opentextbooks/textbooks/30

https://open.umn.edu/opentextbooks/textbooks/71 Readers can go through the Open Access book titled 'Human Resource Management' by accessing the below mentioned URL:

Readers can go through the Open Access book titled 'Organizational Behavior' by accessing the below mentioned URL:

Quick Notes/ Mind Map of the Chapter

Quick Notes/ Mind Map of the Chapter

Your Ideas to add New Dimensions to the Chapter/ Case/ Anecdotes

Section II

Cases and Corporate Anecdotes

2

How to Analyse A Case and Corporate Anecdotes

Sanjeet Kumar Sameer

Associate Professor, School of Agribusiness and Rural Management
Dr. Rajendra Prasad Central Agricultural University, Pusa, Bihar

What is Case Analysis?

Case analysis is basically a real life problem-based learning method. It requires the readers to become active participant in the process and apply their analytical skills including critical skills, reflections, ability to inter-relate multiple aspects of the situations and issues, apply divergent and convergent perspectives in order to identify most feasible solution under constraints and limitations.

How to go about Analysing Case?

Every case and anecdotes have certain core and peripheral issues. Identifying critical incidence is very important to clearly understand the context and problem. Therefore, the participants need to be very active and involved when reading and analysing the cases and anecdotes. The goal is to follow a structured and logically organized format for analyzing every aspect of the case.

Typical steps for Analysing Cases and Anecdotes

a) Preliminary review of the Case

Here one needs to minutely study the background information including all related data, facts and figures (even if these seem unimportant initially), stakeholders, key protagonists, people involved in the process, their roles and expectations, resources, constraints, opportunities, assumptions, premises, goals, ideal solution people may be expecting etc.

b) Situation analysis

The participant is expected to minutely study the case and anecdotes for understanding the situation analytically. It means that he has to focus on various elements of the situation carefully. One has to identify the key stakeholders, issues and problems, events, their inter-linkages and associations that may among each other.

c) Diagnose problem

Many a times we just focus on symptoms and not on the real problem. A symptom can give only temporally solution. Therefore, efforts need to be made to identify the real problem. For this the reader should try to link issues with the evidence. These sets of evidence could be obvious as well as tacit and by linking various small clues one need to figure out the likely problem. This is the core of any case analysis and therefore till the time one is really sure that the right set of problem has been identified, readers need to re-read the case carefully.

d) Listing the problem(s)

Once there is sufficient clarity regarding problems, the same needs to be clearly written so that it is easy to focus on the solution. A well defined problem statement helps in identifying a targeted and focused solution. These problems can be categorised in multiple ways. For example, these can be segregated into immediate problem and long term problem, minor or major problem, critical and superficial problem, urgent and important problem etc. This kind of categorization helps in looking at the identified problems from multiple perspectives.

e) List down the constraints

Readers need to minutely list down various constraints that can affect decision making process. These constraints could be in terms of resources, time, specific stakeholders, internal and external factors, policies, local and community issues, organizational issues, government involvement etc. Also one has to asses these constraints in terms of their significance, ability to control and manage. Some constraints may actually be limitations therefore the readers need to filter out limitations and constraints. This makes the process of decision making bit easier.

f) Develop possible alternatives/ solutions

The readers need to list down all possible alternatives and solutions considering the both the constraints and limitations. These solutions need to be developed based on the nature and category of the problem as highlighted above i.e. urgent or important problem, major or minor problem etc. This focused approach will help in further evaluating the alternatives/ solutions more carefully.

g) Define decision criteria

Before evaluating the alternatives or solutions, there is requirement of having a well defined decision criteria based on the analysis of the case. The decision criteria provides objectivity to the decision making process. Here, consultative approach is advisable as the decision criteria needs to be robust enough to help readers to find out the most feasible solution. It is important to highlight here is that an ideal solution may not be the most feasible solution for a particular context. Although one should attempt to identify the ideal solution also as it helps in assessing the possible consequences if not implemented, yet implementation of the solution is the key. So, many a times feasibility get preference over ideal type.

h) Evaluate alternatives and selection of solution

Each alterative needs to be evaluated for pros and cons, advantages and disadvantages, opportunities and threats in line with the decision criteria. Based on this meticulous evaluation, the alternatives need to be ranked in terms of their suitability under the context. The criteria of suitability can be defined in terms of decision criteria and possible weightage one may like to assign or these criteria well defined in advance in terms of importance. The alternative(s) that rank high on these parameters need to be selected for final implementation.

i) Implement the solution

Once the solution has been identified, there needs to be well documented implementation plan for the identified solution. This stage is very critical as many a times the identified solutions if not implemented do not yield desirable results because of poor implementation design and/or actual implementation. Focus on practical issues help in designing an implementable solution. Also it is important to clearly mention the roles and responsibilities of every individual involved in implementation of the solution, how implementation issues will be resolved, in case any troubleshooting is needed how it will be done.

j) Develop contingency plan

Many a times the solution identified for implementation does not get the desired results. Efforts are made to customize the solution but in certain cases there is a need to implement the next course of action immediately. These sets of alternatives need to be chalked out at the time of implementing the selected solution only so that aligned contingency plan can be immediately implemented due to any issues including goof-ups.

Although monitoring and evaluation is not the integral part of case analysis, it is important for practitioners to have a mechanism of monitoring and evaluation as well so that the solutions are effectively implemented.

Quick Notes/ Mind Map of the Chapter

Your Ideas to add New Dimensions to the Chapter/ Case/ Anecdotes

3

The Mentored or Unmentored

Sanjeet Kumar Sameer

Associate Professor, School of Agribusiness and Rural Management
Dr. Rajendra Prasad Central Agricultural University, Pusa, Bihar

Introduction

After working for more than 10 years in set of private companies, Mr. A K Aggarwal joined a very large government organisation, ITN Ltd (ITNL) (refer to ***Exhibit 1***). He, along with other two freshly recruited MBA graduates, was posted to a project location in the remote district of India after a brief stint at ITNL's corporate office. The top management posted them with a lot of expectation to implement one of the most critical projects of the company. In the first few weeks only, he observed a tussle between the field team and project manager Mr. Amaresh Sharma. This tussle was impeding the project progress. Being new in the organisation, Mr Aggarwal was hesitant to approach any body and was clueless about how to handle the situation. This was also a new situation for Mr. Sharma in his 20 years of ITNL career and he was facing difficulty in breaking this logjam.

New Business of ITNL

It was 2007 when a very large energy company ITN Ltd (ITNL) was thinking to expand beyond its regular oil & gas business. A few options emerged after series of brainstorming sessions attended by senior executives under guidance of renowned strategic consultants. Renewable energy was identified as a potential area for expanding business and therefore Biodiesel group was created by the company to spearhead this business.

Mr. B.N. Mishra, Chief General Manager of ITNL, was made head of this group. He was known for his dynamism and had varied experience of planning, strategy, co-ordination, sales and marketing. A small team was constituted after sourcing manpower from refineries, marketing and pipeline verticals of the company. Most of these officers were selected based on their requirement to relocate to HQ. These three verticals, although part of the same company, had different work culture and market environment. While marketing department was more target oriented, refinery and pipeline focussed more on the Standard Operating Procedures (SOP).

Government of India (GoI) announced a Bio-diesel Purchase Policy in 2006 for blending of up to 5% bio-diesel in diesel. Non edible tree borne oilseeds such as Jatropha, Pongamia, Neem etc. were identified as potential feedstocks for making biodiesel. In response to the policy of central government, many state governments also released their biofuel policy in order to attract investments.

This was an ambitious programme of the government and all energy companies were expected to make effort to realise this target. As per government norms, a Memorandum of Understanding (MoU) was signed between GoI and the Public Sector Undertakings (PSUs) and achievement on this parameter was included under their annual performance rating.

Government promoted plantation of these energy crops on wasteland (government, community, private) only. Lands currently being used for other crops were not to be utilised for commercial plantation of these trees. Accordingly, a team from ITNL visited various potential states of western, central western, central, eastern & southern part of India to submit their project proposals and requested for land allotment. Finally, after series of follow ups by Managing Director (MD), Mr. Mishra and his team, the govt of a state in western part of India allotted 5,000 acres of land to ITNL.

Building Human Capital: The First Impression

Since energy crop plantation required specialised knowledge of agriculture, a proposal was put up to recruit candidates from premier business schools for further carrying out the biodiesel project related activities. Eight MBAs and two experienced candidates were recruited for the job in February 2007.

On request of Mr. Mishra, these 8 MBAs were provided one week corporate induction training at ITNL's training centre. The officers were thrilled to get this opportunity as they had heard a lot about great facilities of this centre. Other mid career enhancement programme was also going on at this centre for officers with 10-25 years of work experience.

While the new joinees were enjoying at swimming pool, some of these mid level officers also came there. Being very enthusiastic and inquisitive to know more about the organisation, four of these MBAs approached them. While discussing about the company, these mid level officers shared that this is the best time of their tenure in the organisation. Once they are posted to any location, they will actually see and experience real ITNL. Some of these officers who were stagnating at the same level (Deputy Manager/Manager) for last 8-10 years shared their bad experiences of the organisation with them. They talked about how charge sheets are issued and disciplinary actions taken against officers even in cases when the officer is just following the instructions of superiors. They told these young officers to be very careful while executing any work otherwise they may end up having similar experience. They added that the career growth in this organisation is dependent on bosses but cautioned that the bosses are not permanent and are not going to remain always there for their rescue. These interactions left a strong imprint on these newly inducted officers.

Mentorship Programme

Mr. Mishra, an engineer by qualification, had learnt about mentorship programme and wanted to implement the same for these new MBAs. He believed that through this programme, the new joinees may be provided sufficient handholding for effectively carrying out the job. A pairing of mentor-mentee was done by him based on availability of mid level officers having 15 plus years of work experience in the organisation. Although he tried to make it a structured one, but due to less monitoring, it ended up being more informal. The mentors were from the same Biodiesel group and did not have past experience of such programmes. Mr. Mishra had advised the mentors to help these officers in quickly adjusting to the new role.

While joining the department, the young officers expected to have their cabin/ workstation available from day one. However, due to logistical constraints, they were not allocated permanent seats/cabins and workstation. The mentor mentee interaction were not frequent and whenever there was such occasion, the mentees would talk to mentors about administrative difficulties they were facing in the office and requesting help for resolving the same. Although the mentors followed with the relevant department, it took good time to arrange facilities viz. seats, PCs etc for the new officers. Mentors would also reiterate to these officers a need to quickly make transition from college to workplace. Since the group was small and a lot of project activities were already lined up, the officers would not get spare time to interact frequently with mentors and other experienced officers. Also, this mentoring programme actively continued for less than a month because of project deadlines. The officers had just started to frankly communicate their issues with their mentors and know more about the organisation.

Mr. Mishra was of the opinion that the new initiative of corporate induction and mentorship programme would have created positive impact on these officers and they would be fully charged up to take up the assignment. Due to his busy schedule and active involvement in this pioneer project of ITNL, he could not find time to take feedback from the officers regarding effectiveness of these programmes.

The Field Team

It was the month of March 2007, when Mr. Mishra called up an emergency meeting with his other junior colleagues Mr. Gaurav. Mittal, Mr. Anirudh

Bagchi and Mr. Amaresh Sharma, who were at middle management level in organisational hierarchy. A decision was to be taken regarding posting of these newly recruited officers. Mr. Mittal, Mr. Bagchi and Mr. Sharma were from different divisions of ITNL and had different experiences and considerations.

Finally after a lot of discussions, 2 fresh MBAs (Mr. Samridh Kumar & Mr. V Nagarjun) and one experienced officer, Mr. Agarwal were identified for being posted in one of the remote districts of Western state of India. Mr. Aggarwal had more than 10 years of experience in private sector. Having worked in such organisations, he was more focussed on result. After knowing about their posting location, the two young officers started collecting information about the district. This district of Western state of India had predominantly tribal population. There were limited power supply and facilities of health & education. Drinking water was loaded with heavy metals. This district was water deficient and agriculture was mostly rainfed. There was issue related to safety as well and therefore people used to avoid late evening / night travel. In past, many cases of looting, snatching and beating were experienced by people staying there. This poor district had very high annual budget allocation and many NGOs were operational in the region. The three officers, after knowing the condition of the posting location were sceptical about their stay in the district.

In the meantime, Mr. Amaresh Sharma was made in charge of the project by Mr. Mishra and these 3 officers were asked to report to him. Other officers were positioned at Head Quarter (HQ).

Mr. Sharma was a target oriented officer. In his past assignments as sales manager, he had been able to exceed targets most of the time and was also awarded best performer of the year award. He was able to get the job done through his own means. His subordinates initially perceived him as less receptive to the new ideas and suggestions of his subordinates. This project was very crucial for both Mr. Mishra and Mr. Sharma as the same was already a part of government MoU, and was monitored by the secretariat of MD.

Going Forward and Backward Towards Achieving the Milestones

The project team was assessing the requirements for timely execution of the project. Bio-energy crop plantation of Jatropha, Pongamia, Neem etc. requires good quality saplings, timely preparation of site, availability of agri-inputs, adequate moisture in soil etc. To streamline project implementation, the HQ had devised a manual for all these plantation activities and the field team was expected to use as a guideline.

Since availability of all resources was to be ensured before onset of monsoon in mid June, these officers worked very hard for next 3 months. They were facing challenges related to their stay and regularly communicating the same to the HQ. However, having a bureaucratic set up, it was taking a lot of time to arrange facilities such as project vehicle, accommodation lease etc. Although they were disheartened by the effort of HQ to streamline facilities for them, they continued their extra effort for success of the project. They would complete survey of the field and shortlist the area based on predefined criteria for plantation activity. Co-ordination with other agencies such as state government, revenue department, forest department, NGOs, community organisations, gram panchayats etc was done by these officers to speed up land acquisition process and lining up various agencies for sapling supply and plantation activity.

There were many NGOs operating in the district and doing similar activity of sapling supply, plantation and agri-input supply, but on small scale. A detailed capability assessment of all organisations operating in the district could not be done due to time and logistical constraints. Since time was less to meet the requirement of this large scale commercial plantation, Mr.Aggarwal who was senior most member of the team posted at project location, suggested to propose one agency on nomination basis for all these jobs (refer to ***Exhibit 2***). For doing so, he needed to justify in the proposal that other agency cannot do the job and the reason for nomination of this agency must be mentioned in the proposal. Further, ITNL did not have any experience of working with this agency.

In government settings, making any statement without having proper basis could have lent him as well as other two officers in trouble. Other organisations and activists might also have sought uncomfortable information from the company using provisions of recently notified RTI Act. Since time was less, he was getting repeated verbal direction from Mr. Sharma to move proposal in this format only. He was advised to prepare a survey report that could support nomination of this agency. The field team had limited exposure and interaction with other experienced employees of ITNL; therefore they were less known to systems and procedures of the organisation. This report and proposal was required to be signed by Kumar and Naragjun as well. Mr. Aggarwal, therefore shared the suggested way forward of Mr. Sharma to these two officers as well. Nagarjun was okay with whatever Mr. Aggarwal decided. Suddenly Kumar remembered what he was told by a few mid level officers during his induction training- "do not go by mere verbal instruction of your bosses, you may be charge sheeted by those only who have given you direction for the same". Since he had some exposure of working in government organisations during his internship, he objected to this idea of Mr. Aggarwal and requested to either get written communication from HQ for submitting proposal in this format or follow general principles of tendering. An alternative way of getting approval of this methodology from senior management was also suggested by Kumar. This would require co-ordination with other functions such as finance, legal, contracts etc and would have been a time consuming process. Hearing all these discussions both Nagarjun and Mr. Aggarwal were also doubtful of the suggestions made by Mr. Sharma.

Any confrontation and stalemate at this stage would have delayed the whole plantation activity. Further, if the crucial monsoon period of 20-30 days was lost, the project would have been delayed by a year. This might have led to non implementation of the dream project thereby causing embarrassment to this highly admired company due to poor performance. This might have led to negative impact on career growth of senior officials associated with the project.

The Perturbed Duos

Being field in-charge, Mr. Aggarwal was unable to know how to respond to this situation. He had discussed the issue with Mr. Sharma but it did not work. He was not sure if he should directly approach Mr. Mishra about the issue. In past, Mr. Sharma was facing difficulty in counselling the team and agreeing them to his idea. Knowing Mr. Mishra as a tough boss, he was also hesitant to approach him for resolution of this issue. He was also considering option to involve Mr. Bagchi and Mr. Mittal, who were mentors of Kumar and Nagarjun respectively, but doubted their effectiveness in resolving the issue? While Mr. Aggarwal and Mr. Sharma were contemplating the next step, they received an email from Mr. Mishra to organise project review meeting next day.

Exhibit 1: Company Profile of ITN Ltd (ITNL)

ITNL was among top five Power Company of India. It is one of the leading government owned enterprises in India. For many years it received excellent performance rating from the government. It had more than 20,000 employees. The operations of the company were spread over all parts of India. It also had foreign subsidiaries in other countries of Asia.

The company had traditionally controlled the energy market in India. After 2000, the company witnessed competition from many public and private companies for market share. Its revenue was declining. To remain in the market, it had to forego the margins. Many of its business associates started switching over to other market players in search of more commission.

Around 2005, its top management was exploring possible avenues of business growth in related segments. The company had identified top three consultants for this exercise and was expecting concrete suggestions on future areas of growth.

Exhibit 2: Definition of terms

Tendering: It is a process of making a formal written offer to carry out work, supply goods etc. Tendering process is aimed at maintaining transparency in the process of obtaining competitive offers. There are three types of tenders: Public, Limited and Single (nomination basis).

In case of public tender, the offers are sought from all organisations meeting certain prequalification criteria. By following this method, maximum participation and transparency in the process is ensured. This may also lead to most competitive offers in case of larger participation. It takes around 3-4 weeks to obtain offers in case there are no issues.

In case of limited tender, a few agencies (generally more than 5) are shortlisted based on secondary sources and approved prequalification criteria. Offers are then sought from these agencies. It takes around 2-3 weeks.

In case of single tender or selecting vendor/contractor/agency on nomination basis, one party/agency is identified for the job. The agency is generally identified based on criteria such as proprietary nature of the job, the specialised nature of job, high reputation of the organisation and its past experience of executing the job efficiently, any exigency which cannot be resolved by other organisation.

Practice Questions

1. List down the concepts of OB and HRM that can are relevant for the case?
2. How these concepts can be applied to understand various events and situations described in the case?
3. How these concepts are interrelated in the context of the case?
4. Applying these concepts, analyse various situations and events described in the case.
5. Read the case carefully and identify major issues and problems. Propose possible solution to the problem by applying the relevant concepts of OB & HRM.

Quick Notes/ Mind Map of the Chapter

Your Ideas to add New Dimensions to the Chapter/ Case/ Anecdotes

4

Coaching & Mentoring Millennials

Pinaki Chakladar

Director, IBM India Pvt. Ltd. Organizational Learning and Change Management Consultant, IBM-certified in 'Digital Change' and 'Organizational Learning

Abstract

Millennials (also known as Generation Y) are the generation that immediately follows Generation X – people who were born in the 80s and 90s and onwards. While Millennial characteristics vary by geography, this generation is ordinarily marked by a significant use and familiarity with social media, communications, and digital technologies. This generation essentially grew up on a liberal philosophy to politics and economics; and therefore, are guided by an open-minded, experimentalist approach at the workplace. These are also the people who are hugely ambitious and want to accomplish much in a short span of time.

Now consider the fact that there are millions of Millennials who are either on the cusp of entering, or are already a part of the workforce in most businesses. Consequently, their attitudes, behaviors and actions are likely to make or break the fortunes of organizations. Needless to say, the way this generation is coached and mentored to develop and retain them needs to reflect their work-and-life expectations. This article seeks to review numerous studies in the 'Coaching and Mentoring' domain relating to impact on Millennials, and the evolution of existing approaches to manage millennials leading to higher organizational performance impacts.

Expectation from Work and Career

Millennials do not distinguish work from life; for them, both work and life blend into a single entity, and therefore, they do not seek a balance between them unlike previous generations (Meister and Willyerd, 2010). This is precisely why they seek work that has a strong focus on personal fulfilment. Their perception of work is that it is a source of opportunities to strike friendships, learn new skills, and associate with a larger purpose. They derive job satisfaction from this singular sense of purpose – they are also the most socially conscious generation since the 1960s. This social consciousness lends them a sense of validation of their purpose – they wish to see their work making the necessary social impact.

On the other hand, they do not see themselves tied to one organization for life – for them; the organization stays relevant as long as their personal goals are met. They value loyalty to the organization only till they see a connect to their sense of purpose and fulfilment. They do not expect long associations with organizations, and are used to working on their CVs since they were practically toddlers.

However, the millennials also want a personalized road map defined for success, and they expect their organizations to provide them with it. If organizations are not careful, they will end up draining the energy of managers in grooming this generation for leadership roles.

Finally, the millennials have corporate empathy and social responsibility deeply rooted within them. What this really means is that they look out for employers who have always backed social causes and issues – they connect with this value deeply, because they believe that only organizations that have social causes at their heart genuinely carry the best interests of their employees and the global community in mind.

Expectation from Their Boss and Organization

From Boss

- Helps me explore my career path in the organization
- Provides me with direct feedback, in a straightforward manner
- Coaches and mentors me
- Sponsors me for personal development programs
- Comfortable with a flexible work schedule
- Recommends me to senior leadership

From Organization

- Develops and enhances my skills to get me future-ready
- Support a strong value-oriented culture
- Offers benefits/reward package with customizable options
- Permits me to blend work with the rest of my life
- Lends clarify on my career path

Things they wish to learn

- Develop functional / technical skills in my domain of expertise
- Develop Self-management and personal productivity

- Develop Leadership skills
- Build Industry expertise
- Inculcate innovation and creativity

Approaches to Coaching Millennials

Keeping the work and life expectations of Millennials in mind, there are several things to consider when coaching them:

Create the right performance environment – It is very important to set clear performance expectations. While it's true that Millennials are free-thinking people, they can also thrive within set performance boundaries, if they are offered a rationale that has good clarity and is logical. The Leadership needs to understand that it is important to be fair and consistent about these performance expectations. When the conversation revolves around development, and not mere monthly or annual conversation, it will certainly strike a chord with them.

Millennials just want to be clear about the fact that they are constantly learning and growing with each experience. This is because they view development as a top priority and, as such, view ongoing feedback from their Manager/Coach on how they are doing. Therefore, the Coach needs to provide the feedback at all times, regardless of whether it is positive, constructive or neutral in tone. All that the Millennials want is to engage in meaningful conversations and give/take feedback via two-way conversations – because they do not want to be told what to do; they can figure that out by themselves.

Offer effective guidance and leadership – Millennials always look up to their managers and hope to derive learning from them. They also aspire to model their behaviors on the leaders they can look up to, those who value them as individuals and value their contributions. Therefore, it's important to let them know how important they are to the overall success of the organization. The right leadership for this generation would comprise people with coaching skills, good listening skills, and the ability to connect one-on-one to clarify expectations and drive results through accountability. Only such leaders can deliver high-quality guidance for this generation of people.

Offer Work that is meaningful – For Millennials, the 'purpose' is more important than the salary they draw. Millennials want to know how the work they perform has a true impact on the communities around them, and the society at large. Only that kind of work will interest them that contributes to the vision of the organization. Millennials are curious to know that the work that they do truly matters. Therefore, while delegating responsibilities to them, it is important to show how the responsibility they are being assigned contributes to the larger goal or mission for the organization.

Nurture and Treasure their unique talents – Millennials are defined by a "Go-getter" attitude; they are ready to conquer whatever challenges confront them. They are not scared to stand up for what they believe is right and seek management support utilize their talent for the betterment of the entire organization. Therefore, it is important to encourage them to nurture and leverage their unique talents and "go-getter attitude to accomplish the loftiest goals.

Create an environment of collaboration – Teamwork is another attribute that characterizes the millennial generation – they truly thrive while working with others towards common goals. They have been programmed for success and they often have to set their own career roadmap to accomplish their goals. Consequently, it is important to prompt them on how they plan to accomplish these goals and then back them in their efforts.

Kinds of Mentoring for Millennials

Mentoring of millennials serves a high degree of proximal outcomes, such as career and psychosocial functions, apart from role modelling needs. Findings from research (Meister and Willyerd, 2010) have led to the broad identification of three kinds of mentoring that will prepare Millennials for success without requiring experienced managers to spend all their time coaching them. While these approaches will also work with other generations, they are especially effective with Millennials, because they suit their mobile, collaborative lifestyle and need for immediacy.

Reverse Mentoring

This mentoring approach is an innovative one, where it puts the responsibility for organizing mentoring to the junior employees, who learn from senior executives by mentoring them (Murphy, Wendy Marcinkus, 2012). A Millennial is matched and tagged to a manager and assigned to educate him on things that are their area of strength – for example, how to use social media to connect with customers. This then becomes an effective way for young employees to gain a peek into the higher echelons of the organization – by the time their mentees retire; the younger generation would have had a better understanding of the organization's business.

The purpose of this arrangement is primarily knowledge sharing, where the emphasis is on the mentee learning from the mentor's subject expertise or technical expertise and gaining from the generational perspective of the millennials. Additionally, the mentors stand to gain in terms of leadership development. This form of mentoring is contextualized in the mentoring literature as an alternative form of mentoring, with unique characteristics and support functions exchanged that distinguish it from other developmental relationships.

Of course, there will be times when the older workers will offer guidance and feedback to the younger workers – so, in that sense, the mentoring becomes mutual. The young employees get to benefit from having a potentially accelerated career track, as the mentoring arrangement offers their profiles as references among senior executives of the organization. On the other hand, the executive mentees have the opportunity to gain understanding of this new, irreverential segment of their workforce they might not have otherwise known. This form of mentoring can also enhance an organization's ability to share with employees honest, timely, and useful coaching – most importantly, the millennials place a high premium on having a manager who "will give straight feedback."

There are certain characteristics that cuts across generations, though. Like employees from other generations, Millennials too want to feel valued, empowered, and engaged at work. As a matter of fact, this is a fundamental need, not an issue with a specific generation. Millennials also pose one question constantly - "Am I continuing to learn and grow?" The way organizations respond to that question may help retain its competitive advantage by way of retaining the talent from this aspirational generation of millennials.

Group or Peer Mentoring

This form of mentoring targets groups of millennials, and therefore, is slightly less-resource-intensive in nature. It is still an effective way of providing Millennials with instant, ongoing feedback, which they crave so much (Grant-Vallone, Elisa J; Ensher, Ellen A., 2000). This initiative could be driven by a senior manager or by peers – in both cases, the organization could set up a technology collaboration platform that allows employees to define mentoring in their own terms – the downside here, though, could be that group think could influence individual preferences.

Social collaboration platform such as these could enable employees to share their knowledge and insights with other colleagues through byte-sized podcasts and video clippings and discussion threads, apart from conventional training resources. The young workforce could view content and rate learning nuggets according to their relevance, depth and quality of content. What's more, they can connect with the experts who posted the learning nuggets in case they wish to eke out additional information about the topics covered by the learning nuggets.

Organizations are no strangers to Peer mentoring – it is becoming increasingly commonplace and has been found to be effective way to facilitate knowledge creation and sharing (Bryant, Scott E., 2005). Every organization values organizational knowledge creation and sharing, since it is an important source of competitive advantage. Results from the research (Bryant, Scott E., 2005) suggested that a peer mentor training course increased perceived levels of peer mentor knowledge and skills. Results also indicated that higher perceived levels of peer mentoring were related to higher perceived levels of knowledge creation and sharing.

Finally, studies indicate that protégés' satisfaction with a formal peer mentoring program included the extent to which mentoring behaviors met career-related needs and psychosocial needs and the amount of time spent with the peer mentor. Results indicated that the degree of career and psychosocial functions served by a peer mentor were strongly related to protégés' satisfaction with the mentoring relationship.

Anonymous Mentoring

Anonymous mentoring is a recent development and a unique model to connect this generation with the mentors, without having to worry about someone trying to impress you on something or you trying to impress someone else or be apprehensive about what you write will be used against you. Internet technologies allow you that anonymity, which truly liberates both the learner and mentor, freeing them up to explore all kinds of topics under the sun, exhorting them to get to the root cause of chronic organizational problems.

This mentoring method leverages psychological assessment and a background review to match mentees with seasoned and skilled mentors outside the organization (Meister & Willyerd, 2010). The mentor-mentee transactions could be conducted entirely using online tools, with both the mentee and the mentor (usually a seasoned coach or executive) remaining anonymous. This type of mentoring engagement is generally funded by the mentee's company, and lasts between six to 12 months.

The good part about this form of mentoring is that while Millennials have elevated expectations of mentoring as a process, they also have great willingness to try new mentoring models consisting of anonymous, and even multiple mentors.

Usually, the coaches can't imagine themselves running anonymous mentoring exercises, their efficacy being looked at with much scepticism. But once this mentoring method takes off and matches the coach / mentor with a mentee using the psychological assessments, the coach is pleasantly surprised at how well they were paired. It turns out to be a highly intimate relationship while remaining completely anonymous.

As a matter of fact, when the mentoring arrangement draws to a close, both the mentor and the mentee feel as though they were losing a close friend. Many of them vouch for the fact that anonymity of this arrangement was an unexpected boon. Contrast this with how they felt at the beginning - Initially, both of them would have thought it to be an awkward mentoring arrangement, but later, they both could see why the anonymity was required.

Mentoring with Micro feedback

This is a mentoring tool that addresses the crying need for instant guidance with minimal resources. You can imagine this as performance assessment for the FB and Twitter addicts — concise, precise and almost real time.

Practice Questions

1. In this Chapter you studied about coaching and mentoring. Based on the concepts discussed, try to identify situations around you and develop a case let/ story/ anecdote that seem relevant to you.
2. Please go through the articles mentioned under further readings section and analyse which aspect of the articles can be relevant in the context of coaching and mentoring.
3. What other concepts of OB and HRM can be associated with coaching and mentoring?
4. How these concepts are interrelated?
5. Applying these concepts, analyse various situations and events described in the case let/ story that you have developed.

Further Readings and References

Ekta, K. & Sameer, S.K. (accepted for publication), "Changing dynamics of work: Interplay of organizational and spousal support, person-job fit, work-life balance and work-family integration", ABS International Journal of Management (ISSN: 2319-684X)

Meister, J. C., & Willyerd, K. (2010). Mentoring Millennials. Harvard Business Review, 88, 68-72.

Sameer, S.K. & Priyadarshi, P. (2019). A Framework for managing internal employability by leveraging motivating job characteristics- Role of promotion and prevention focused job crafting. Paper presented in GLOGIFT 19-Flexibility, Innovation and Sustainable Business, an Annual International conference of the Global Institute of Flexible Systems Management (GIFT), organized by Indian Institute of Technology (IIT), Roorkee during 6-8th December 2019.

Sameer, S.K. & Priyadarshi, P. (2020). Interplay of organizational identification, regulatory focused job crafting and job satisfaction in management of emerging job demands: evidence from public sector enterprises. International Review of Public Administration, 26(1),73-91, https://doi.org/10.1080/12294659.2020.1848024

Sameer, S.K. & Priyadarshi, P. (2021). Role of Big Five personality traits in regulatory-focused job crafting. South Asian Journal of Business Studies, 10(3), 377-395, https://doi.org/10.1108/SAJBS-03-2020-0060

Sameer, S.K. & Priyadarshi, P. (2022). Interplay of Job Characteristics, Promotion- and Prevention-Focused Job Crafting and Internal Employability. In: Anbanandam, R., Rangnekar, S. (eds) Flexibility, Innovation, and Sustainable Business. Flexible Systems Management. Springer, Singapore.

Sameer, S.K. (2022). Managing employees' performance in Indian public sector undertakings during COVID-19 pandemic-induced blended working. South Asian Journal of Business Studies, Vol. ahead-of-print No. ahead-of-print. https://doi.org/10.1108/SAJBS-05-2021-0170

Sameer, S.K. (2022). The interplay of digitalization, organizational support, workforce agility and task performance in a blended working environment: Evidence from Indian public sector organizations. Asian Business & Management. https://doi.org/10.1057/s41291-022-00205-2

Sameer, S.K. and Priyadarshi, P. (2023), "Regulatory-focused job crafting, person-job fit and internal employability–examining interrelationship and underlying mechanism", Evidence-based HRM, Vol. 11 No. 2, pp. 125-142. https://doi.org/10.1108/EBHRM-08-2021-0163

Sameer, S.K. and Priyadarshi, P. (2023). "How emotionally intelligent employees manage their internal employability through role-based job crafting? – evidence from public sector enterprises, International Review of Public Administration, Vol. 28 No. 3, pp. 265-287, DOI: 10.1080/12294659.2023.2256099

Quick Notes/ Mind Map of the Chapter

Your Ideas to add New Dimensions to the Chapter/ Case/ Anecdotes

5

The Expatriate Vs Local Supervisor Dilemma

Abhishek Tiwary

Senior Vice President and Global Head-HR (BPS) for Tech Mahindra

Background

Alpha Partners is a large professional services partnership firm comprising over 180,000 professionals with presence in more than 156 countries around the world. It primarily deals in three main businesses – Statutory Audit of listed companies, Taxation Services (both corporate and individuals& Advisory services to leading private companies and governments around the world.

In India, the company has a total of approximately 20000 employees and partners divided between two broad entities – Alpha Partners India (AI) and Alpha Partners Global Services (AGS). AI is the primary client service firm in India and is engaged in professional services to clients in the Indian geography. AGS, on the other hand is a joint venture between Alpha Partners firms in U.S, U.K. and India – all three holding equal shares of AGS. Formed in 2011 and structured exactly like any other member firm of the Alpha Partners network, AGS provides Advisory, Tax, and Audit professional services to 80+ member firms worldwide within the Alpha Partners network. It score delivery responsibilities are pointed inward and therefore can also be categorised as a captive center It is a strategic, global delivery organization that engages with Alpha Partners member firms around the world to provide innovative, scalable and customized solutions. Simply put, AGS is a captive off shoring, shared services center for Alpha Partners. The company has other centers around the world serving internal customers.

For the last four years, Alpha Partners has ranked as an 'Ideal Employer' on a global index of the world's most attractive employers competing with the likes of Google and Microsoft. The index is based on the opinions of close to 130,000 students from renowned academic institutions across many economies, including the United States, China, Germany, France, the United Kingdom and India. Alpha Partners provides its employees with award-winning diversity initiatives and a formal mentoring program.

Since its inception, AGS has been focused on making member firms increasingly relevant to their clients. By leveraging the experience and talent of more than 7,000 professionals in India, AGS has succeeded in offering its stakeholders incremental value quickly and efficiently.

AGS has grown significantly in the first five years since its inception. The growth has come through both organic and inorganic means. About half of the company's employees belong to acquired companies. The average age of the employees of the company is 25 years and the average age of managers is 28 years. By all means, it can be called a 'young' company. In the first few years of setting up, the company employed a number of expatriates from various countries for knowledge transfer and training purposes.

The expatriates were expected to groom young local talent and prepare them to deliver quality service to US, UK and various offices of Alpha Partners around the world. In a way, they directly supervised the work of the young employees of AGS. The company was careful in choosing and selecting their best employees from across the world for the AGS postings as it was critical for AGS to be successful. A number of employees applied for few vacancies in AGS India and chosen few came for assignments. At some point, the company had plans to replace all expatriates with local management.

Evolving Relationships

The expatriates at once liked the young, vibrant and dynamic work place in India. The first few batches of expatriates and their local teams were responsible to lay foundation stones and get the local Indian operations of AGS running. At that time, the relationship between the expatriate supervisors and their local Indian subordinates was easy and collaborative. As the AGS operations matured over the years and the local teams got better control and understanding of the overall business and operations, the local supervisor's role visibly began to enhance and correspondingly the expatriates' role began to decline. Their roles were far more distanced from core operations and reduced to just managing relationships with their partnership.

The local team members were supervisors of junior staff and some of them were showing great early leadership traits. In fact, these supervisors sometimes proved to be better than their expatriate managers. The junior staff liked working with them. The performance management of junior staff was also done by these local supervisors, therefore the relationship grew much deeper. Over a sustained period of time. the local supervisors also grew a credible relationship with their onshore partners almost rendering the expatriates redundant. This was the beginning of an interesting work dynamics between expatriates who believed they owned the business and local supervisors who felt the business ideally belonged to them as they were hands on managing the business. This marked the beginning of a deep people issue that AGS had to quickly get rid of, since it had soon started showing in the performance of the organization.

The Decline Performance and the Induced Stress

The overall decline in the relationship at senior levels within the organization induced a stress in the system which was felt by the junior most employees. The key metrices used to measure business success started to steadily fall. The onshore teams began to think twice before sending work. This resulted in lower increments and bonuses of the group. Many employees began to leave as a result.

The Proposed Solution

As soon as the financial year ended, the onshore Partners responsible for sending work to India sat to discuss the underperforming AGS. A number of possible root causes were discussed.

- A group of Partners believed that the issue lied in selection of the right expatriates for the AGS role. They believed that there had to be change in criteria of selection of these employees as the organization had evolved
- A second set believed that the expatriates needed to be pulled back completely as AGS had now matured to a level where local supervisors could do the job with direct oversight from the business leaders onshore.
- A third set of recommendation was to put the two sets of supervisors through a learning intervention on collaboration.

The Final Solution

A combination of all three proposed solutions was adopted over a period of 18-24 months.

- As a first step, all the supervisors were put through rigorous training on collaboration and team work immediately following the meeting
- Based on the inputs from training and the evolved stage of AGS, the company changed the criteria of selecting expats. The job description of the expatriates was also re-written to suit the needs of an evolved AGS organization structure.
- Finally, the company decided to gradually reduce the number of expatriates over a period of 24 months and only kept some key and critical roles in AGS reserved for expatriates. This allowed for better and more robust local control and better discipline in running of the business.

Practice Questions

1. Critically analyse the solution offered by the author. Do you agree?
2. List down the other concepts of OB and HRM that can are relevant for the case?
3. How these concepts can be applied to understand various events and situations described in the case? Please use multiple perspectives and approach.
4. How these concepts are interrelated in the context of the case?
5. Applying these concepts, analyse various situations and events described in the case.
6. Read the case carefully and identify major issues and problems. Propose possible solution to the problem by applying the relevant concepts of OB & HRM.

Further Readings

Ekta, K. & Sameer, S.K. (2023), "Changing dynamics of work: Interplay of organizational and spousal support, person-job fit, work-life balance and work-family integration", ABS International Journal of Management (ISSN: 2319-684X)

Sameer, S.K. & Priyadarshi, P. (2021). Role of Big Five personality traits in regulatory-focused job crafting. South Asian Journal of Business Studies, 10(3), 377-395, https://doi.org/10.1108/SAJBS-03-2020-0060

Sameer, S.K. (2022). The interplay of digitalization, organizational support, workforce agility and task performance in a blended working environment: Evidence from Indian public sector organizations. Asian Business & Management. https://doi.org/10.1057/s41291-022-00205-2

Quick Notes/ Mind Map of the Chapter

Your Ideas to add New Dimensions to the Chapter/ Case/ Anecdotes

6

Employee Entitlements & Remuneration in International Subsidiary of Public Sector Company of India: A Choice Between Consistency and Differentiation

Sanjeet Kumar Sameer

Associate Professor, School of Agribusiness and Rural Management
Dr. Rajendra Prasad Central Agricultural University, Pusa, Bihar

Introduction

BOGL had set up its wholly owned subsidiary company Bikas Oil Inc. (BOI), Moscow, Russia. Mr. Atanu Saxena and Mr. Samish Batra were recently selected for being deputed to BOI. Being a foreign location, this posting was considered as monetarily rewarding and good for career growth of employees. Subsidiaries of other oil and gas companies were also present in Moscow as this location was considered as the oil and gas business hub of Russia.

BOGL has limited number of subsidiary companies outside India and approximately 15 officers have been posted in these foreign subsidiaries. This number was very small considering more than 10,000 officers in BOGL. In some subsidiary, it had also hired local contractual employees. The company had adopted the policy of foreign posting almost 25 years ago when Indian economy was getting liberalised during early 1990s. This policy document was one of its kinds for the company and did not have any blue print and was untested. The posting policy covered eligibility criteria of posting; entitlements & remunerations for posted employees along with certain restrictions etc. It has occasionally been updated in the past. Some of its guidelines had generic reference to the policy of Ministry of External Affairs for its Indian Foreign Service (IFS) employees.

The two BOGL officers Mr. Saxena and Mr. Batra joined BOI, Moscow in April 2016. Their compensation had two components – home salary and foreign component (allowances) along with other perks and benefits. With a lot of effort, their salary processing was regularised, both from BOGL and BOI for respective components. Moscow also had other subsidiaries of Indian PSU. These BOI officers came in contact of the employees of other foreign subsidiaries. They also came to know about their remuneration, perks and benefits.

In the meantime, BOGL had approved a remuneration policy for its employees deputed to its recently formed foreign joint venture company in Brazil. This policy was made considering policy of other joint venture partners' policy of posting in order to minimize the variations. The three officers also had access to this policy document.

After interacting with their counterpart in Brazil and employees of other PSU subsidiaries at Moscow, the three officers compared this policy with their existing policy and found some beneficial provisions were missing in their policy. Also some of the attractive provisions of their existing policy were missing in those of other companies or BOGL's subsidiary in Brazil. They discussed those beneficial missing provisions with BOGL head office in India and submitted a proposal for incorporating those provisions in the existing policy.

Mr. Ramesh Jha was co-ordinating with BOI office from India. He was told to make a comparison of policy of foreign posting at BOI, BOGL's subsidiary at Brazil and subsidiary of other PSU in Moscow. A detailed comparison was made and submitted to the senior management. A lot of variation was found in terms of home salary calculation methodology, foreign allowances, perks and benefits etc. While existing policy at BOI was found to be monetarily more rewarding, the prevailing guidelines for Brazil based subsidiary of BOGL or other PSUs subsidiaries had more employee friendly provisions such as frequent travel to home country, location based allowances etc.

BOGL's senior management reviewed the proposal submitted by these officers and found that they had discussed only those provisions which they were not provided. They had not touched upon such provisions where they were paid substantially higher amount such as foreign allowance, allowances for hotel stay etc. This approach by the officers had led to wrong perception about them in the minds of senior management of BOGL.

Since this issue was related to human resource (HR) policy of the company and therefore was required to be decided by the HR department. A mid level officer of the department Mr. Nishant Sharma was directly approached by these three officers. Mr. Sharma briefed Mr. Chandan Ahluwalia, Head of Corporate HR, about the proposal. Interestingly, when the foreign posting policy of BOGL was approved 25 years ago, Mr. Ahluwalia was a member of the policy drafting team. Any change in the posting policy at this moment would have a long term implications for the company. Mr. Ahluwalia was contemplating about his next course of action.

Practice Questions

1. List down the concepts of OB and HRM that can are relevant for the case?
2. How these concepts can be applied to understand various events and situations described in the case?
3. How these concepts are interrelated in the context of the case?
4. Applying these concepts, analyse various situations and events described in the case.
5. Read the case carefully and identify major issues and problems. Propose possible solution to the problem by applying the relevant concepts of OB & HRM.

Quick Notes/ Mind Map of the Chapter

Your Ideas to add New Dimensions to the Chapter/ Case/ Anecdotes

7

The Curious Case of Culture Transformation At AXYZ Ltd.

Pinaki Chakladar

Director, IBM India Pvt. Ltd. Organizational Learning and Change Management Consultant, IBM-certified in 'Digital Change' and 'Organizational Learning

"Unseasonal thundershowers," remarked Mr. Ritesh Prasad, as he peered out of the windows at the overcast sky and the storm raging outside – the landscape outside was no different than the one raging inside AXYZ ltd, a family-owned, logistics firm. Mr. Ritesh Prasad, or RP as he was popularly known, had been the CHRO at AXYZ for a decade now. He was a personal favourite of Mr. Pulkit Sahni, the founder-owner of AXYZ, and was quite popular with all levels of hierarchy, because of his genial disposition and general sense of empathy towards the firm's workforce.

He had seen many executives come and go, with their share of ups and downs, but never had he witnessed the churn the organization was faced with now, ever since the new CEO, Mr. Anil Chatrath, had taken charge of the operations of the firm a few months back.

It all began with the 2020 envisioning set by the new incumbent CEO, who believed AXYZ was long overdue for an overhaul in its operations. He believed that AXYZ was still living in its past glory, and it badly needed a transformation to move from its cultural origins as a family business to that of a modern enterprise that was capable of seamlessly conducting operations worldwide.

To set this in context, AXYZ was primarily a first-generation, family owned business that had a paternalistic culture, founded on assumptions that emphasized personal and charismatic characteristics of the founder and family. Headquartered at Delhi, its operations were spread across four more cities of India – Rajkot, Mumbai, Hyderabad and Chennai *(Refer Appendix 1.2)*.

After half a decade of stagnating growth, diminishing profitability and plummeting customer satisfaction, the founder-owner, Mr. Pulkit Sahni, decided it was time for a shakeup – the first step in that direction was to hire a competent CEO, who could usher in a professional culture in the organization, placing it on the next growth spiral. Another reason for Mr. Sahni to hire an outsider as the CEO was to get an experienced professional to mentor his children and prepare the second-generation leaders from the family to lead the company in future.

The Case for Transformation: Opportunities and Challenges

Enter Mr. Anil Chatrath, an executive with a high leadership pedigree and proven credentials in driving business transformations. Mr. Chatrath set about his business at AXYZ Ltd in a clock-work fashion. In order to ring in a professional culture in this firm, he rolled out Vision 2020 – establishing individual initiative, risk-taking, accountability and impersonal rules as the means of getting work done.

However, there were quite a few stumbling blocks that confronted Mr. Chatrath while embarking on this cultural transformation. For starters, the company had a paternalistic pattern of culture, where relationships were organized hierarchically. The leaders were traditionally family members – the CEO was the son of the owner, while the COO was his niece – who retained all power and authority and made all the key decisions. Most of the management positions in the company were staffed by people who were hardcore family loyalists.

The leaders in the family distrusted outsiders and closely supervised the employees. Furthermore, the family members were accorded preferential treatment, and the employees were supposed to carry out the family's orders without questioning them.

Mr Chatrath's Vision 2020 envisioned an organization where the employees' focus would be on individual achievement and career advancement. This vision entailed a Performance Management System that rewarded employees on their ability to contribute to the profits of the business. Additionally, it trimmed the owning family's involvement in the business, placing the emphasis on professional management.

The intent behind designing the new cultural blueprint was to infuse individual initiative, creativity and innovation, besides improving efficiency. Employees were incentivized to do their jobs quickly and efficiently. Mr. Chatrath was in a reactive mode, however – being a professional manager, who was brought into this family firm to turn it around, he was forced to put out fires, take a proactive stance in cutting costs by instituting "modern" management techniques and bring in overall accountability across the firm.

And sure enough, his Vision 2020 articulated all of that – he started by instituting a performance Management system that focused on individual initiative, personal and team accountability. It included a Rewards system that discouraged nepotism and celebrated excellence. He followed this up by restructuring the company, laying off low-performing employees, to send out a hard-hitting message – shape up or ship out. *(The new organization structure is referenced in the basic business organogram in Appendix-1.1)*

He also replaced a majority of the previous management with his own team and encouraged the new managers to compete with one another in carrying out his strategic vision and programs. Bonuses and other incentives were offered to those individuals who succeeded, penalizing those who didn't.

Efficiency and cost control were the watchwords of the Chatrath regime. A Change Management team was constituted to sell the advantages of the new professional culture – it communicated to the rank and file of the organization about the benefits that new, innovative ideas and professional management techniques would bring to the firm. Vision 2020 was planned to improve the firm's accounting, marketing, and other operating systems and make the business run more efficiently. Implicitly, the new management wanted to reduce the firm's ties to the past, thereby enabling all to see the new possibilities that was purported to move the firm in new directions.

Unfortunately for the new CEO, the major weakness of the professional culture instituted by him was that it alienated the employees who were used to working for the family under a different set of assumptions. The impersonal nature of the programs created immense discontent and ill will among union members and led to bitter fallout among the top management team members – it was a standoff between two groups of people – the family loyalists and the professionals hired by the new CEO. Thus, the professional culture that the CEO ushered in at AXYZ Ltd proved to be short-lived.

In the months that followed, things went from bad to worse. It led to growing absenteeism, turnover, and unhealthy competition among individuals and among departments. To add to it, there was low morale, and low commitment among not just the management teams, but among the rank and file of the firm.

To the firm owner, Mr. Sahni, this was no less than a nightmare – all his plans to transform the firm had backfired, despite hiring the best management staff, holding the right intent, putting elaborate Change plans in place and constituting a team of Change champions to set the plans into motion.

He immediately summoned Mr. Ritesh Prasad (the CHRO) to take a quick stock of the events, facilitate a board meeting to plan for corrective actions before things slid into a downward spiral even further.

So what went wrong with the Change initiative?

First, the founder-leaders of the family firm underestimated the impact of their existing assumptions and behaviour on the firm – when the new CEO announced Vision 2020, they did not apprise him fully of the various nuances and assumptions around which the family firm had grown – for example, a web of personal relationships to drive business in each function; informal work processes that operated on trust basis; focus on group instead of individual endeavour and excellence.

In order to initiate the new culture, therefore, the CEO should have unlocked and disconfirmed *(Lewin, 1952, Schein, 1985)* aspects of existing work culture before making any modifications to that – this is something the CEO failed to do.

Second, a complete overhaul of existing work culture is generally difficult to accomplish, and therefore, a hybrid model could have been targeted. Vision 2020 could have retained some of the core values of the family business and blended it with the primary tenets of the Professional culture the CEO was aspiring for. Therein, the new Vision could have adopted one of the primary tenets advocated by Edgar Schein (Schein, 1985) - to promote individuals who share most of the basic values and assumptions of the leaders but vary on one or two dimensions. These individuals are similar to the leader in terms of general attributes, so they can gain acceptance. However, they are also different enough to introduce change.

As an example, the new crop of Professional managers could borrow the key assumptions of the leader of the family firm, with the exception of the assumption about human nature – these kinds of hybrid org-cultural models could have eased the firm into the new cultural landscape that is more participative and acceptable.

Third, while the CEO replaced the top management (comprising family members) with Professional Managers, with different assumptions, in order to change the culture more quickly, what he didn't comprehend was the inherent disadvantages of such an approach – this approach compromised the existing rituals, routines, and practices that had helped the firm in the past.

Additionally, long-term employees, who were ignored for promotion as a consequence of the new Performance Management system, soon became disgruntled and demoralized, and casted a negative influence on the rest of the workforce.

Problems with ensuring Continuity in the Firm

Most leaders of family firms are faced with a common question at one point of time or the other in their career: How to change the culture of their business to make it more effective? This is the same question that confronted Mr. Pulkit Sahni, when he wanted to transform AXYZ.

However, he failed to answer this question in his attempt to accelerate the transformation at his firm. The answer really lay in the fact that he should have first changed his own assumptions and behaviors that shaped his firm.

One of the actions he could have possibly taken was to analyse his firm's culture, list the key assumptions of his leaders, and then planned for change, through interventions by outsiders; for example, consulting firms.

Analyzing the Culture and Planning for Change

To conduct the analysis, he could have formulated a joint-action team, comprising members of the organization as well as consultants hired from outside. The insiders could have then assisted the consultants to understand the nuances associated with the firm's culture, while the outsiders could document the more tacit dimensions of the culture — by way of observations and often by asking "stupid" questions about surprises they encounter.

The joint-action team could have then interviewed individuals and observed behaviours across various levels and locations throughout the firm. Both insiders and outsiders could conduct the interviews to drive collaborative inquiry, documenting what appear to its members to be the significant artifacts of the culture, then uncover the rituals and routines of the culture that make up the perspectives shared by members of the organization.

For example - Who is rewarded and who is penalized here? How are decisions made here? What kind of people are hired? The team could also try to uncover subcultures— groups holding different beliefs—by asking interviewees which individuals or groups hold opposing views. This is because, at times, the problems in family firms are the result of a set of subcultures at conflict - such as family members versus nonfamily members.

Conclusion

Culture change at AXYZ Ltd was not easy, and to have executed any of the initiatives that the new CEO planned also required the total commitment of the top leadership, since the leaders largely create and shape the cultural patterns of their business. In case of XYZ, while Mr. Sahni had initiated the change and hired the CEO to drive it, he himself kept away from this mammoth exercise. He should have understood the effects of his firm's culture and take steps to ensure it fostered the desired cultural patterns that would allow the business to thrive and grow.

To understand and manage the opportunities present in family business cultures is not easy, and it is not practiced at many family businesses, but it is necessary for the leaders who wish to ensure the continuity of their businesses and the well-being of their families in the long run.

Appendix 1.1: Basic Organogram of AXYZ Ltd.

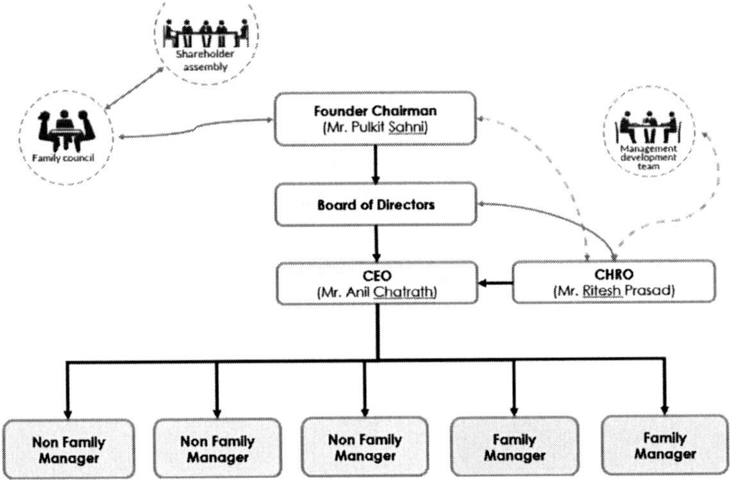

Appendix 1.2: Illustration of India-wide Logistics operations of AXYZ Ltd.)

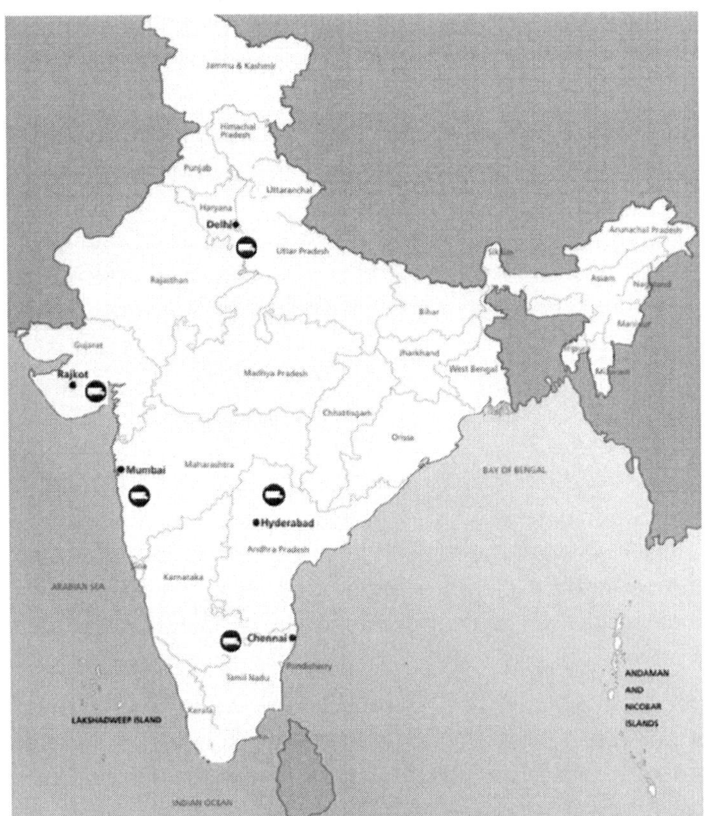

Practice Questions

1. In addition to the concepts explained above in the case, please list down the other concepts of OB and HRM that can are relevant for the case?
2. How these concepts can be applied to understand various events and situations described in the case?
3. How these concepts are interrelated in the context of the case?
4. Applying these concepts, analyse various situations and events described in the case.
5. Read the case carefully and identify major issues and problems. Propose possible solution to the problem by applying the relevant concepts of OB & HRM.
6. Based on this case, develop a case let in not more than 3 pages depicting similar situation and dilemma.
7. What additional perspective you can add in analysing the case?

References and Further Readings

Schein, E. H. "Organizational Culture and Leadership: A Dynamic View." San Francisco: Jossey-Bass, 1985.

Lewin, K. "Group Decision and Social Change," in Readings in Social Psychology, rev. ed., eds. G.E. Swanson, T.N. Newcomb, and E.L. Hartley (New York: Holt, 1952).

Sameer, S.K. (2022). The interplay of digitalization, organizational support, workforce agility and task performance in a blended working environment: Evidence from Indian public sector organizations. Asian Business & Management. https://doi.org/10.1057/s41291-022-00205-2

Sameer, S.K. (2022). Managing employees' performance in Indian public sector undertakings during COVID-19 pandemic-induced blended working. South Asian Journal of Business Studies, Vol. ahead-of-print No. ahead-of-print. https://doi.org/10.1108/SAJBS-05-2021-0170

Quick Notes/ Mind Map of the Chapter

Your Ideas to add New Dimensions to the Chapter/ Case/ Anecdotes

8

When in Rome Act Like Romans Is it True in Case of Parent Company & Its International Subsidiary Relationship?

Sanjeet Kumar Sameer

Associate Professor, School of Agribusiness and Rural Management
Dr. Rajendra Prasad Central Agricultural University, Pusa, Bihar

Introduction

Mr. Ramesh Gupta was transferred from a well established business division to an emerging division of one of the largest companies of India, TN Limited (TNL) (refer Exhibit-1). Mr. Gupta had more than 30 years of work experience in TNL when he joined this new division. The division was primarily into acquisition and management of oil and gas assets in India and abroad. Recently the company had acquired a very large asset in North America. To manage this asset, a Houston, USA based subsidiary of TNL named as BM Limited (BML) was created (refer ***Exhibit-2***). Considering scale of business and complexity involved, the company decided to post some of its brightest officers there. Through competitive and transparent process of selection, three officers viz. Mr. Birendra Sharma, Mr. Rajat Lal and Mr. Ashok Agarwal were selected for being posted at Houston, USA.

Mr. Sharma was known for his diligence and was highly systematic. His subordinates and seniors both considered him a very competent officer and senior management including Mr. Gupta had posed high faith in him. This was his first posting outside TNL's head office since he joined the organisation 10 years back. Mr. Rajat Lal was good at networking & had strong financial skills.

Before leaving for Houston, one of the important steps for these officers was issuance of work visa. After series of follow ups and co-ordinated effort with immigration agency in Houston, finally they received their work visas. Delay in visa processing was also linked to bureaucratic culture of TNL which necessitated obtaining multiple approvals from various competent authorities.

Mr. Sharma and two other colleagues were advised to immediately leave for Houston. These officers left for their posting location without taking their family along. They thought to take back their family after they make necessary accommodation arrangements for them.

Barely 2 months had passed and they decided to visit India to take back their family. The policy of parent company permitted reimbursement of travel expenses on account of personal reasons only once during their posting tenure apart from their joining and repatriation time. Thus, if reimbursements of expenses are made for these officers this time, they will have to spend money from their own pocket in case they have to visit India subsequently during their period of posting. This was quite surprising for these officers. Although they had gone through details of entitlements of foreign posting but somehow they missed this aspect.

The three officers requested senior management of TNL to provide dispensation against this provision as they were advised by Mr. Gupta to leave immediately for Houston. This claim was not taken supportively by Mr. Gupta. While efforts were made to resolve this issue, it was taking longer time than expected by the officers. The three officers felt that TNL is not supporting them sufficiently even though this was not their mistake. Finally, after much effort and follow ups for more than a month, their request was approved. This episode had negative impression on Mr. Sharma and his colleagues about supportive stand of parent company. TNL's senior officers had opposite opinion and they felt that the approval could not have been received had they not followed up with various HoDs aggressively.

By the end of 3 months, these three officers had already begun to get acclimatised at Houston. During this period they were able to line up their accommodation and develop good networking with other project partners and local agencies. They started learning culture of this place.

USA in general has a low context culture. People to people interactions are based on direct communication. The replies are generally frank even between bosses and subordinates. In India, subordinates are expected to be able to have ability to identify needs of bosses and subtly try to fulfil them.

BML was headed by Mr. Birendra Sharma as Country Manager. Directors on the board of this subsidiary company were from senior management grade of TNL. Although Mr. Sinha was in middle management cadre of TNL, this responsibility of heading a company had impact on his behaviour. The USA based project was coordinated from India by another mid level officer Mr. Ramesh Agarwal of TNL who was senior to Mr. Sharma.

A partners' meeting was scheduled at Houston which was required to be attended by Mr. Gupta. Earlier, when Mr. Gupta used to visit Houston in relation to project matters, he had to be sure about his own accommodation/ travelling arrangements. Now since there was an office of the subsidiary and officers were posted there, he too had expected all arrangements to be made by BML. As a norm, when senior personnel of TNL visit any office/location in India, the arrangements are made by the office being visited. This has become part of company's norm and culture.

When Mr. Gupta landed at Houston airport, he was expecting somebody from the office to receive him. The three officers posted at Houston had observed during their interaction with other partners that this type culture is not promoted in the USA. Further, such requirement was not communicated to them by TNL. Therefore, they did not visit the airport to receive Mr. Gupta. At the end of this two day visit, Mr. Gupta did not feel great about the visit.

On his return to India, Mr. Gupta reviewed various activities done by the subsidiary company. He stressed upon cost control and advised reduction in non-project expenditure and office related expenses.

The subsidiary company depended completely on TNL for any approval. As the office was recently set up, a lot of proposals on various issues were being put up for approval by senior management of TNL. On many instances, their proposal was either delayed or rejected. Further, interaction between these officers and TNL senior management gradually reduced and it was limited to only official matters. These all issues lead to development of lack of trust between TNL and BML officers. In India, subordinates and bosses confide in each other on their personal issues as well.

This hostile environment was not liked by anybody associated with the project. Neither some of the members of senior management of TNL nor officers posted in Houston were willing to end this hostility. And time and again the officers were advised to start dialogue to break this mistrust. But they were reluctant.

It was the season of performance appraisal in TNL. As per performance evaluation mapping, the final appraisal of the officers posted abroad was to be done by senior management of TNL. One of the appraisers for Mr. Sharma was Mr. Gupta. Amidst this hostile situation, the ratings of these three officers were not up to their expectation. Officers claimed that their evaluation was done subjectively without giving due importance to the good work they had done in setting up new office and related infrastructure, supporting mechanism, deployment of manpower etc. in a culturally different country.

Lack of trust and amicable environment reduces project performance and the same was also experienced in this case. Earlier it was anticipated that these officers will work effectively and bring best global practices for implementation at TNL as well the subsidiary company BML. Apart from Mr. Sharma, other two officers were also not able to bring any positive change. Due to this misunderstanding, the team output was declining. Further, other departments were gossiping about these developments which were further creating negative perception about the foreign subsidiary and parent department.

In the meantime, one of these three officers Mr. Lal had to visit India in connection with his urgent personal work. He was also planning to visit TNL's head office. Being project co-ordinator, Mr. Agarwal was contemplating means to sort out the problem and improve relationship between the parent company TNL and the foreign subsidiary BML as this issue may be asked by the senior management anytime. With a lot of thoughts in his mind he was exploring how best he could utilise visit of Mr. Lal for this purpose. Suddenly Mr. Agarwal got a call from Director of TNL. Perplexedly, he moved towards Director's Cabin thinking what all questions Director may ask him and their possible replies.

Exhibit-1: About TN Limited (TNL)

TNL is among top 10 Energy Companies of India. It is a public enterprise of the Government of India and has more than 80% direct and indirect shareholdings. TNL is more than 50 years old company and has recently grown into diversified businesses viz. alternative energy, upstream oil and gas business, power etc. Being a government undertaking, it has a bureaucratic structure. Hierarchy has its own importance in this company.

Exhibit-2: About BM Limited (BML)

BML was created as a wholly owned subsidiary of TNL in order to manage projects in North America region such as USA. TNL has strategically opened its office at Houston, as this is a major oil and gas business hub in North America. BML was projected as face of oil and gas business of TNL. The parent company wanted this subsidiary to be a place of best global practices and expected BML to translate these practices into the parent company as well. The culture of this company was to be developed same as other companies operating in this part of the world.

Practice Questions

1. List down the concepts of OB & HRM that can are relevant for the case?
2. How these concepts can be applied to understand various events and situations described in the case?
3. How these concepts are interrelated in the context of the case?
4. Applying these concepts, analyse various situations and events described in the case.
5. Read the case carefully and identify major issues and problems. Propose possible solution to the problem by applying the relevant concepts of OB & HRM.

Quick Notes/ Mind Map of the Chapter

Your Ideas to add New Dimensions to the Chapter/ Case/ Anecdotes

9

The Cultural Divide

Abhishek Tiwary

Senior Vice President and Global Head-HR (BPS) for Tech Mahindra

Context

Value Services Incis (VSI) incorporated and headquartered in New York, USA and with presence in 14 countries across the globe. With a total of 14000 employees globally, about 23% of their workforce is based in India in their offices across Mumbai, Delhi and Bangalore. Their India headquarters is in Mumbai.

The company's business is primarily in research services. They pride themselves in providing high quality and in-depth research output across all domains, industries and geographies. They serve 40% of the fortune 500 companies as their clients and by 2020, they aim to cover 70% of the list.

Last year, they surpassed their annual revenue target - their global annual revenue stood at USD 3.2 Bn. India business however contributes just about 6% of the topline. The VSI business in India is divided into two broad parts – a profit center (VSI India) and a cost center (VSI Business Services). The employee split between the two businesses is about 30% and 70% respectively. Represented in India as two separate entities, both the businesses are 100% owned subsidiaries of the VSI US (the parent company).

The profit center in India caters to Indian clients and their subsidiaries and affiliates in other countries. The cost center part of the India business caters to back office operations supporting other countries' offices of VSI - primarily USA. They also offer shared services for many support services of VSI Inc.

The Cultural Divide

The client facing profit center and the back office businesses have constantly nurtured subtle undercurrents of disharmony – one always trying to outdo/ overpower the other, in the eyes of the management in the US. The local leadership of the two entities though always optically displayed good harmony and collaboration, the discontentment with each other's presence was pretty obvious both internally as well as externally. The Leadership of the client facing firm felt that they should have full ownership of the back office and be responsible for it. However, the management of the company was never convinced about that decision. The back office on the other hand always showed its allegiance to the US Management, their rightful owners. The situation between the two organizations was anything but normal. The leadership had a huge trust deficit that percolated down to the staff of the two organizations.

VSI in India enjoys a good brand in the talent market and is sought after by many organizations. Many join the firm thinking that they could switch between the front office and the back office per their personal situations and aspirations alternating between organizations that allowed careers in a global environment (the back office) as well as an opportunity to work with clients in the Indian marketplace (the front office).

The VSI headquarter in the USA was worried about the situation on the ground in India. While India was a profitable business and one of their largest operations worldwide, such cultural divide did not help while the company was on a journey to unify under one brand and collaboration was key in the agenda. The new CEO, John Lucas had a clear mandate – to create a deep sense of culture and one-ness of the brand in minds of the clients in the company. He really wanted to create and leverage the power of the brand and had clearly instructed his leadership to ensure it was followed in letter and spirit.

The COO and the second in command of the company, Harry Bergit had a task at hand. He needed to straighten things in India. He knew if he did not act now, the brand in India would take a solid beating. He had lately been reading some painful comments about the company in some online forums – especially about culture of distrust in India. There was only one person, he thought who could do this.

The Case of Venkat Naidu

Venkat was part of the core leadership at New York that oversaw India operations. He was hired by Harry about six years ago when they were starting India operations. Venkat's Indian origin was also a reason for VSI to hire him. Within VSI (and more importantly) in India, Venkat enjoyed a lot of respect and camaraderie in both the back and front offices. In fact, he had a role to play in hiring at least 30% of the current leadership on both sides, especially at the time when the team was setting up. He spent a large amount of time in India in the first twelve months of setting-up of India operations. So in a nutshell, he was no stranger to the trust deficit that India leadership of both sides experienced today.

Harry convinced Venkat to take up the role of 'Culture Advisor' of both the units in India. In this role, he would directly report to Harry in New York. His mandate was very clear - to get the culture right and align it with the overall culture of the organization. The two organizations also needed to collaborate better.

Sitting in the plane to India, Venkat began to think hard of what could have gone wrong. He had known some key individuals and some of them even quite personally. He knew their strengths and their shortcomings. He was wondering if he had made hiring mistakes with some of them. But he wanted to keep his thought process positive. He wanted to get to the right outcome without disturbing the status quo. The India unit was profitable but had seen a decline in profits for the first time last year. That had served as alarm bells for the management.

On arrival in India, Venkat quickly got into action. He met a number of employees both senior and junior, old and new. He also met every member of the leadership team. He also met many ex-employees of VSI to understand why they left. The problem was more deep-rooted than he thought.

Initial Assessment

He penned his first thought to Harry in 6 weeks of coming to India. Here are the excerpts of his report:

1. There was a sharp divide in the cultures of the two organizations. The client facing organizations thought of themselves as being more superior to back office since they were a profit center. Many in front office made the remark to back office employees –

    a. “we work hard to earn you your salaries”

    b. “we are, therefore you are”

2. Many policies of the firm were differently interpreted and (often twisted) within the two organizations. There was no explanation for some to be different. For example – a back office employee would get Rs 250 for a meal if he/she stayed back late in the evening while a front office employee would get Rs 500 for the same meal.
3. Lack of internal career paths – an employee from one organization could seldom move to the other organization even if he/she fit the criteria. The movement within organizations was discouraged by the leadership. As a result, the organization was losing a large number of people to the competition.
4. The leadership of the two organizations did have some great collaborative people but they were pulled back by a vast majority who did not believe in collaboration.

5. Many exited employees spoke of how stifling and frustrating the cultural clash could be. The rate of attrition for the back office was higher because they would often compare themselves with the front office employees who they believed were treated better than them.

6. Venkat also made a very interesting observation about the role of the leadership in New York in creating this divide. This included Venkat himself – something he did not realize the impact of, sitting thousands of miles away from the ground. This, probably was his most important observation – not to say that the others were any weaker argument for the cultural divide. While setting.

    a. When the back office was being set up, the then front office leadership was hinted that it was their own child and they needed to run it. But somewhere down the line, they did not happen. Many India Office leaders who had helped set up the back office grudgingly moved on.

    b. Whenever the leadership members came to India, they would disproportionately spend more time with the back office and not spend much time with front office and its employees. This further created stress amongst the front office staff.

Conclusion

Venkat now believed he had a good understanding of the problem at hand. He now had to work on the action plan. One thing he was sure of. There were three protagonists in the play – the two India entities and the head office. He knew the head office in New York had to play their part in sorting this problem out. And to that extent, he now understood why Harry chose him for the job – he could work with both India and New York teams with equal ease.

What should Venkat now do to get some immediate, mid-term and long-term results?

Practice Questions

1. List down the other concepts of OB and HRM that can are relevant for the case?

2. How these concepts can be applied to understand various events and situations described in the case? Please use multiple perspectives and approach.

3. How these concepts are interrelated in the context of the case?

4. Applying these concepts, analyse various situations and events described in the case.

5. Read the case carefully and identify major issues and problems. Propose possible solution to the problem by applying the relevant concepts of OB & HRM.

Further Readings

Ekta, K. & Sameer, S.K. (2023), "Changing dynamics of work: Interplay of organizational and spousal support, person-job fit, work-life balance and work-family integration", ABS International Journal of Management (ISSN: 2319-684X)

Sameer, S.K. & Priyadarshi, P. (2021). Role of Big Five personality traits in regulatory-focused job crafting. South Asian Journal of Business Studies, 10(3), 377-395, https://doi.org/10.1108/SAJBS-03-2020-0060

Sameer, S.K. (2022). The interplay of digitalization, organizational support, workforce agility and task performance in a blended working environment: Evidence from Indian public sector organizations. Asian Business & Management. https://doi.org/10.1057/s41291-022-00205-2

Quick Notes/ Mind Map of the Chapter

Your Ideas to add New Dimensions to the Chapter/ Case/ Anecdotes

10

Am I a Performance Manager or A Mentor?

Abhishek Tiwary

Senior Vice President and Global Head-HR (BPS) for Tech Mahindra

Background

GMPK is a large professional services partnership firm comprising over 155,000 professionals with presence in more than 156 countries, providing Audit, Tax & Advisory services to some leading companies and governments around the world. In India, the company has two entities with a total of approximately 20000 employees and partners.

For the last four years, GMPK has ranked as an 'Ideal Employer' on a global index of the world's most attractive employers. The index is based on the opinions of close to 130,000 students from renowned academic institutions across many economies, including the United States, China, Germany, France, the United Kingdom and India. GMPK provides its employees with award-winning diversity initiatives and a formal mentoring program.

The GMPK Global Services (GGS) is one of the two entities in India. It is a strategic, global delivery organization that engages with GMPK member firms around the world to provide innovative, scalable and customized solutions. Simply put, GGS is a captive offshoring center for GMPK. They have other centers around the world serving internal customers.

Set up in 2008, GGS is a joint venture between GMPK U.S., GMPK U.K., and GMPK India, along with the support of GMPK International. KGS provides Advisory, Tax, and Audit professional services to 80+ member firms worldwide.

Since its inception, GGS has been focused on making member firms increasingly relevant to their clients. By leveraging the experience and talent of more than 7,000 professionals in India, GGS has succeeded in offering its stakeholders incremental value quickly and efficiently.

GGS has grown significantly in first five years. The growth has come by both organic and inorganic means. About half of the company's employees belong to acquired companies. The average age of the company is 25 years and the average age of managers is 28 years. By all means, it can be called a young company. In the first few years of setting up, the company employed a number of expatriates from various countries for knowledge transfer and training purposes.

The expatriates were expected to groom young talent and prepare them to deliver quality service to US, UK and various offices of GMPK around the world. In a way, they directly supervised the work of the young employees of GGS. The company was careful in choosing and selecting their best employees from across the world for the GGS postings as it was critical for GGS to be successful. A number of employees applied for few vacancies in GGS India and chosen few came for assignments.

The role of expatriates in the company

The company depended a great deal on the expatriates as they acted as knowledge repositories and were tasked with ensuring transfer of knowledge to local supervisors who were expected to lead the delivery out of India as the expatriates headed back to their respective countries.

The job description of the expatriate was designed largely by their respective home countries. Sometimes, the job descriptions would differ for two individuals from different countries as they had been written by two individuals at two different times in two locations.

However, the most common aspects of their role expectation inter alia included

1. Keeping the purpose and profitability of their own home country in mind first always and every time.
2. Manage performance of their team members in order to achieve the purpose set for them by their home countries.
3. The delivery efficiency of the GKS was a joint responsibility of the expatriates with the local Indian management of the company.

After passing through a stringent selection process, the individual is assigned the responsibility of working in India. Since GKS is a strategic asset for the GMPK, there is much fanfare for someone who gets selected to this roleas a stint with GKS is considered to be in the domain of high performers.

Culture

The company has a truly diverse portfolio with a hundreds of products in markets around the globe. GMPK has always endeavoured to provide its employees an inclusive environment where employees and management coming from all walks of life are meant to perceive similar experience.

Case of John Roberts

John was high performing Senior Manager in the Manchester office of GMPK and had been employed by the company for almost nine years. He has been amongst the top 20% employees of the Audit division. In the last performances appraisal discussion, his supervisor advised him to avail an opportunity in an overseas office of GMPK if he wanted to be considered for the position of a Director in future. Amongst various option GKS came out to be the top choice.

Based on the discussion with his Supervisor, John applied for the position of Senior Manager and got selected to be a part of the GKS team in India.

Arrival

As John boarded the plane at Heathrow airport in London along with his wife, Rubina and the 6 yrs old daughter, Lucy, he was excited a slightly nervous. His life had been fairly organized and straight forward up until now and this assignment to India seemed like the first major deviation. He was wondering if he had taken a big risk. His wife had to quit her plush job in an investment banking firm to accompany him and explore the eastern side of the world. Lucy was enthusiastic and amused about the trip as her friends had been telling her that she would find many snakes in the country and she had never seen one. Also, She had very fond memories of her nanny for four years, Promila who would tell her enchanting tails of India.

John arrived in India in accordance with the policies and reported to Alan Butch who incidentally also came from the UK office. Alan was a Partner and leading the Audit operations of GS which primarily provided offshored services back to the GMPK UK office.

The Team

John was assigned a team of 45 professionals who were equally enthusiastic about their new manager. The rumor mills were abuzz with how easy of difficult he would be as a manager. The average age of the team was 24 years with 85% of the team being bachelors. Their previous manager, Bill Henkel was from Leeds in UK and they were sad about him returning. He had built a great rapport with the team and was personally committed to their success and well being. He was very affectionate in his dealings and very knowledgeable at his subject.

John clearly knew he had an immediate uphill task waiting for him. He had to equal or surpass the level of affection and relationship his predecessor had built. The task became even more uphill as the performance appraisal cycle was round the corner and many in the team were expecting a great rating and/

or even a promotion. He knew he had to quick get to know the team and understand their individual performances and make the 'right' decisions on their careers.

Radhika Athawale

Radhika was a bright enthusiastic employee in John's group. She was the seniormost person amongst the group both by tenure and age. She acted as the local supervisor of the group and their go-to person in case of doubt. At the same time, she was also sought when the leader of the group (in this case always the expatriate and now John) who needed Radhika's advice when it came to matter of team management or any local nuances.

Radhika provided mature advice and was always a top performer in the group. In the first meeting with John, he was assured of both her maturity and her stature in the team. He knew she could help him become successful pretty quickly. Radhika was a manager and was only a level below John. They both had same kind of experience and number of years although John was still her supervisor.

Radhika on the other hand was apprehensive and had even approached Alan enquiring about John's experience and his credentials. Alan had assured her that he was a great performer and that she would enjoy working with him.

The Year End Appraisal

Radhika and John finished the year end appraisals for all individuals in the team seamlessly. Radhika had a great role to play. She sat with John on all discussions and helped him through some really difficult conversations.

It was now time for Radhika's appraisal.

John had been thinking hard about it. He had even consulted Alan and requested him to sit through it. Radhika on her part had an unusual request to make. She wanted Alan to conduct her performance appraisal. She believed John was more like her peer. She didn't mind working with him on a day to day basis but she believed he couldn't really provide any professional guidance to her beyond a point. Alan was quick to advice her to talk to John.

The Mentoring relationship

Alan reached out to John and asked her not to react negatively to Radhika's request. His only advice to John was to promote himself to be Radhika's mentor rather than just her performance manager. John was confused at this advice and insisted upon Alan to explain a bit more. Alan said that the relationship between Radhika and him had to reach a higher ground where she seeks him out for advice and doesn't merely report to him.

John had to graduate himself gradually to being Radhika's mentor and the process could take weeks/months to get there. Alan also said that a mentoring relationship will ensure not only Alan's success but also Radhika's success as a professional.

John found merit in the advice and started working on a deeper and meaningful relationship with Radhika where he let Radhika operate with the team almost independently, without much interference. And at the same time, he would be there to advice and guide her when she needed it.

Radhika seemed to have found her space and seniority and Alan as a mentor was now her new found guide and advisor. Something she had lacked in all her supervisors in the past.

John on his part, invested time in knowing Radhika – her personal, educational and professional background.

The yearend appraisal discussion was but a small part of the overall relationship that was now stable and solid.

Radhika was now beginning to introspect about her relationships with some of her senior team members. Could she be a mentor too? May be that's the guidance she must seek from John?

Practice Questions

1. List down the concepts of OB and HRM that can are relevant for the case?
2. How these concepts can be applied to understand various events and situations described in the case?
3. How these concepts are interrelated in the context of the case?
4. Applying these concepts, analyse various situations and events described in the case.
5. Read the case carefully and identify major issues and problems. Propose possible solution to the problem by applying the relevant concepts of OB & HRM.

Further Readings

Ekta, K. & Sameer, S.K. (2023), "Changing dynamics of work: Interplay of organizational and spousal support, person-job fit, work-life balance and work-family integration", ABS International Journal of Management (ISSN: 2319-684X)

Sameer, S.K. & Priyadarshi, P. (2021). Role of Big Five personality traits in regulatory-focused job crafting. South Asian Journal of Business Studies, 10(3), 377-395, https://doi.org/10.1108/SAJBS-03-2020-0060

Sameer, S.K. & Priyadarshi, P. (2022). Interplay of Job Characteristics, Promotion- and Prevention-Focused Job Crafting and Internal Employability. In: Anbanandam, R., Rangnekar, S. (eds) Flexibility, Innovation, and Sustainable Business. Flexible Systems Management. Springer, Singapore.

Sameer, S.K. (2022). Managing employees' performance in Indian public sector undertakings during COVID-19 pandemic-induced blended working. South Asian Journal of Business Studies, Vol. ahead-of-print No. ahead-of-print. https://doi.org/10.1108/SAJBS-05-2021-0170

Sameer, S.K. (2022). The interplay of digitalization, organizational support, workforce agility and task performance in a blended working environment: Evidence from Indian public sector organizations. Asian Business & Management. https://doi.org/10.1057/s41291-022-00205-2

Sameer, S.K. and Priyadarshi, P. (2023), "Regulatory-focused job crafting, person-job fit and internal employability–examining interrelationship and underlying mechanism", Evidence-based HRM, Vol. 11 No. 2, pp. 125-142. https://doi.org/10.1108/EBHRM-08-2021-0163

Sameer, S.K. and Priyadarshi, P. (2023). "How emotionally intelligent employees manage their internal employability through role-based job crafting? – evidence from public sector enterprises, International Review of Public Administration, Vol. 28 No. 3, pp. 265-287, DOI: 10.1080/12294659.2023.2256099

Quick Notes/ Mind Map of the Chapter

Quick Notes/ Mind Map of the Chapter

Your Ideas to add New Dimensions to the Chapter/ Case/ Anecdotes

11

Driving Performance Culture A Case of CTN Ltd.

Sanjeet Kumar Sameer

Associate Professor, School of Agribusiness and Rural Management Dr. Rajendra Prasad Central Agricultural University, Pusa, Bihar

Sambit Sharma, Chairman, CTN Limited (CTNL) received a letter from the Chairperson of 3rd Pay Revision Committee (PRC) for Public Sector Undertaking (PSU) of India. The letter inter alia sought feedback on existing Performance Management System (PMS) at CTNL and suggestion for its continuity or modification, if any. In past also, CTNL's suggestions on various human resources (HR) matters had been valued by the Committee and many of those suggestions were retained in their final recommendations to the Government of India. Mr. Sharma, off late, had heard many stories, both positive as well as negative, about the PMS within CTNL and was therefore contemplating to thoroughly review the system. He marked the letter to Ratish Agarwal, Director (HR), CTNL to take necessary action.

CTNL: A Jewel of the Govt of India

CTNL was among top five energy companies of India. Government of India had more than 80% direct and indirect shareholdings in the company. It was more than 70 years old company and had recently grown into diversified businesses viz. alternative energy, upstream oil and gas business etc and formed more than 20 joint ventures with reputed companies from India and abroad to speed up business expansion. The company met the vital energy needs of the consumers in an efficient, economical and environment-friendly manner. It had more than 50,000 customer touch points all over the country, even to the remotest place of Leh and Nicobar.

CTNL had more than 30,000 employees, spearheading the mission of energy security of the nation. Fifty % of these employees were in executive / officer grade (see Exhibit 1 for Grade wise distribution of employees). Government of India looked forward to the company for implementing its ambitious projects. In the past, the company had been able to stand up to the expectations of the government. CTNL was also a major revenue generator for the government. It had paid highest taxes and dividends to the government in recent past.

Human Resource Policy and Performance Management System at CTNL

Since CTNL was a Government company, pay, remuneration, rules for employee management etc., mainly for officers below board level, were defined by the Government. Broad guidelines on the same were approved by the government and like other PSUs; CTNL had to align itself to those guidelines. To prepare those guidelines, government of India had constituted PRC in the past. In 2007, it constituted second PRC for the PSUs which submitted its recommendation in August 2007. The government implemented this report for the PSUs with certain modifications. The report defined policy regarding pay and perks, career progression etc. This also defined methodology of performance management. All PSUs were told to implement a PMS based on bell curve (see Exhibit 2). CTNL implemented the recommendations of the Committee in May 2009.

CTNL engaged a top Consultant in May 2009 for designing the PMS based on government guidelines. The company wanted to have an online portal for the same. The Consultant designed the PMS with following key features:

- PMS had two Modules: Performance Planning and Performance Appraisal
- It consisted of Key Matrices viz. Key Result Areas (KRAs) and Key Performance Indicators (KPIs)
- KRAs were defined into two categories: Mandatory KRAs and Supplementary KRAs
- Mandatory KRAs must cover 50-80% of the overall plan depending upon the level of officers
- Each KPI had stretch assigned to it indicating level of difficulty (1 mean most difficult)
- KPI wise stretch was to be assigned by the Appraiser. Average stretch of plans of subordinates under a supervisor should not exceed 0.95. Stretch signified level of difficulty in completing the task associated with the KPIs. Stretch value ranged from 0 to 1; 1 signified most difficult KPI.
- Goal setting had to be mutually agreed between subordinate, supervisor and reviewer. In case of disagreement, the decision of superiors prevailed.
- Every year, the performance planning exercise was scheduled to be completed during April month.
- Plan was required to be reviewed every 6 month. As per the design of the PMS, feedback should be provided by the supervisor to his subordinate at an interval of 6 months.

- Performance Appraisal was also required to be completed during April month of every year.
- There were four types of ratings: Excellent, Very Good, Good and Poor
- Appraisal had to be done by the appraisee, his boss (appraiser), reviewer and finally the Final Appraisal Authority (FAA).
- Appraisal to be done for KRAs as well as on criteria such as Competence (Co), Values (Va) and Potential (Po) parameter.
- In general, KRAs constituted 60% weightage in the final score. Co,Va and Po had 40% weightage. However, this proportion varied as one move from lower level to higher level.
- Appraisee must agree to the ratings given by the Appraiser before further assessment by reviewer and FAA could be initiated. However, in case of disagreement, the appraiser has the final say.
- Appraisal by Reviewer and FAA did not require agreement by the appraisee.
- Appraisee could appeal against his ratings to the Appellate Authority within the stipulated time.
- Appraisal at all level was to be done in such a way that not more than 20% employee get excellent ratings and not less than 10 % get poor ratings.
- Ratings by reviewer were to form basis of performance bonus payout. Ratings by the FAA were to form basis for career progression.
- Comments of Appraiser, Reviewer and FAA on Co, Va, Po was not to be made visible to the appraisee. Only score on these parameters and total score were visible.
- Additionally for mid level managers, 360 degree feedback system was used wherein feedback from self, peers, subordinates and seniors was to be obtained using 20 item measures. Ratings were made out of 5 (up to one decimal point). Score on 360 degree feedback had 20% weightage in the final score used for career progression.

Initial Days of PMS Implementation

The Consultant made presentation of the PMS to the Board of Directors in the month of July 2009 and suggested steps for its implementation. Emphasis was given on training and workshops. It was suggested to do live demo of the process. A video about the steps involved was also suggested to be made and uploaded on company's intranet for easy access and reference of employees.

Such tools were also implemented by the Consultant in some of the top performing energy companies in the world and found to be successful. Based on these discussions, PMS was finally approved by the Directors and the same was to be implemented by Director (Personnel) in CTNL. Director (Personnel) created a PMS Implementation Cell within Human Resource vertical and deployed Executive Director level officer Mr. Manik Chandra as In charge of the cell. Mr. Chandra was also part of the same team which was co-ordinating with the consultant during PMS design.

The PMS cell spearheaded implementation process of PMS. It sought name of co-ordinators from each department. Functional departments nominated the co-ordinators on adhoc basis without assessing aptitude, interest and desired orientation of the employee. Also majority of departments deputed those employees for this job who were perceived to be less productive within the department. CTNL had more than 100 departments under 6 key verticals. Workshops and trainings along with live demos were organised across departments. HR personnel emphasised on objectivity part of the system. Employees felt that due to objectivity they would now be able to perform according to agreed plan and performance ratings would be based on transparent evaluation process. Majority of employees welcomed the system. Those who have been stagnating at a particular position considered this new system a rescuer due to objectivity in evaluation and transparency.

It was already July 2009 when the PMS was adopted by CTNL. Goal setting was required to be completed within next one month. Being a new system; KRAs, KPIs, reporting structure etc had to be mapped in the system. Sufficient exposure was required to be provided to employees to use the system effectively. Queries of different kinds would reach to HR department and they had to respond to these queries quickly. Since PMS was implemented through coordinators identified for each department, there were some instances when queries could not be addressed quickly. This was mainly due to less time available to coordinators to understand each and every aspect of the new system.

Employees experienced various issues while setting the goal. Some of those issues were related to quantifying the targets, more specifically in those roles which were meant to do exploratory work, scout for businesses, develop systems and procedures etc. For those projects where CTNL did not have role in day to day operation nor had control over project planning and operation by virtue of being minority partners, employees faced a lot of difficulty in correctly setting the targets and associated milestones. Bosses too perceived difficulty in critically evaluating goals set by the subordinate. As KPI stretch of the subordinate was to be fixed by the supervisors, they were experiencing problems in allocating the right stretch. Also, each supervisor had to ensure that average stretch remains 0.95, therefore a tendency of averaging out for each KPIs was observed. Bosses felt comfortable in putting 0.95 stretch for each KPIs. Bosses were not able to discriminate between high stretch and low stretch KPIs.

With a lot of follow ups by Mr. Chandra and Mr. Agarwal , the goal setting exercise for year 2009-10 was completed by the end of September 2009 with a delay of one month from the target date.

Critical Time of Performance Appraisal

It was the month of April 2010, when HR department circulated communication to employees for commencement of performance appraisal. A deadline of May 2010 was fixed for completion of the process across all levels.

Information Systems personnel working with in HR department were asked to keep track on the progress of performance appraisal using log in details. By 15th May, only 10% of employees had initiated self appraisal process. Series of reminders were sent, however even after those efforts only 40% of employee could initiate self appraisal by May 31, 2010. Interestingly, the cycle of appraisal by self, appraiser, reviewer and FAA was completed in case of only 10 % employees. It was found that majority of these employees belonged to HR functions only.

To complete the process, an extension of 7 days was provided. Due to sudden rush, the server could not support simultaneous log-ins. Another extension of 7 days was given with a request to not wait up to the last date to complete the appraisal process. All these efforts could ensure only 75% completion of the appraisal. Still 25% was left. Mr. Chandra briefed Mr. Agarwal about current status and conveyed his efforts to complete the process within stipulated timeline. Agarwal was not satisfied with the progress and was contemplating way out for timely completion. He sought help from Mr. Sambit, Chairman, CTNL.

A meeting was convened by the Chairman on request of Director (Personnel) which was attended by Director (Marketing), Director (Business Development) and Director (Production). These three verticals were the main revenue generator for the company and majority of PMS related pending cases belonged to these verticals only. Mr. Sambit knew that without their support, the appraisal exercise could not be completed within time. During the meeting, it emerged that a disincentive mechanism should be put in place in order to avoid further delay in submission of performance appraisal. Accordingly, it was decided to deduct 10% score from the final rating in case employees do not complete the process within next 7 days. This deduction would mean lower bonus payout and chances of reduced final ratings. Based on the concurrence by three powerful Directors, a directive was issued by Mr. Chandra about penalty associated with delay submission beyond next 7 days.

Now as per new directive, the cycle of self appraisal, appraisal by appraiser, reviewer and FAA was to be completed within next 7 days. Remaining employees were more interested in just completing submission of their appraisal and therefore did not actively engage in the process. Neither feedback was exchanged between appraiser and appraisee. In some cases, citing paucity of time, self appraisal and appraisal by supervisor was done by the appraiser itself. Appraisers appeared to be completing the appraisal exercise casually.

Reactions of Employee about the PMS after Performance Appraisal Results

Appraisees, in general, faced difficulties in effectively using the PMS. Appraisers faced even more difficulties. Since the PMS did not allow excellent rating beyond 20% at each level, appraisers were finding it very difficult to assess performance of each subordinates objectively just based on his/her actual performance. Further rating on Co, Va, Po by reviewer and FAA appeared to be highly subjective. Some employees also called it "appraisal based on face value". In many cases, it was observed that even after score on KRA was very high (excellent category), but score on CoVaPo was so low that total rating got reduced to good category. Many similar cases were observed across the departments.

Further, number of appraisee under an appraiser, a reviewer and a FAA also varied according to the group, projects and department. This ranged from single appraisee under an appraiser to as high as 20 appraisee. Similarly, at reviewer level this number ranged from 5 to 40. At FAA level, numbers of eligible appraisee were 20-400. Generally, the FAA used to evaluate a larger group, and employees had limited opportunity of interaction with his FAA, who was also the final authority for performance rating.

Each department used its own criteria to identify excellent performer by tweaking score on CoVaPo which was very subjective. In many cases, excellent ratings by the appraiser were downgraded to good rating. In some cases, good rating was upgraded to very good and very good was changed to excellent. This was done in order to ensure the percentage limit associated with bell curve.

The system which was initially perceived to be highly objective by employees turned out to be very subjective and prejudiced. Hope of many employees got shattered. Many employees became cynical about the system and found this system a defunct one, infact they considered it to have a damaging effect on the organization. Intra group and intra department rivalry was observed among employees. This had started affecting team performance. Some employees also developed a feeling of dissonance. The PMS was also perceived to be a morale damaging tool. Many FAAs and Reviewers were uncomfortable with mandatorily rating 10% employees under poor category and giving excellent ratings to only 20% employees. This meant even there is a non performing project; one can have 20% employees under excellent ratings. Similarly, in case of a tough project, one has to mandatorily give 10% rating under poor category even after excellent progress of the project. Such instances spread a perception of injustice among employees.

PMS as a Tool for Organizational Culture Development

Annual review of PMS implementation was done by the consultant as per terms of contract signed during designing of the PMS. HR department compiled list of issues and discussed the same with the consultant. Consultant suggested providing more training and exposure to the employees, particularly who were performing the role of appraisers as their inputs were very critical for successful implementation of the system.

The top management of CTNL wanted to develop a healthy competitive culture within organization in order to improve productivity and increase per employee contribution across various departments. This was also required to make CTNL efficient in dealing with market competition. The consultant suggested making successful implementation of PMS as a critical KRA for each Head of Department (HoD) in order to develop this competitive culture by objectively differentiating performance of employees. It was easier said than done. The functional department did not see this task as a part of their role. They considered it as a role of HR department. However, with support of Mr. Sambit, a guideline was rolled out by Mr. Chandra making HoD as prime driver of the PMS.

It was the month of September 2010 when performance payout was to be announced based on PMS ratings. The payout brought happiness among some of excellent rated employees and more dissatisfaction among many employees. For the first time, there was significant difference in the payout among employees. This difference was more pronounced in case of middle and senior management level employees.

This created a lot of dissatisfaction about the PMS and employees wanted to know the comments mentioned by their appraiser, reviewer and FAA under Co, Va and Po component and also general comment on their overall performance. Top management was reluctant to make these comments visible to the employees as they thought this will further create dissatisfaction among the employees. This may also increase rivalry among team members and create an unhealthy competitive environment.

Already some group of employees of other PSU had filed a petition in Delhi High Court regarding non-transparent system of PMS implemented in their organization after implementation of 2nd pay Committee. The Court gave its verdict on the case in the month of October 2010. It directed all PSUs to make annual performance appraisal report visible to the employees from immediate effect. CTNL had no option but to implement this order. The annual performance appraisal report was made visible to all employees in the month of December 2010. The portal was open for 15 days.

Employees had a mixed reaction at the disclosure of their final annual performance appraisal report. Now they were able to see comments of their appraiser, reviewer and FAA. In many cases, while downgrading score of the appraisee, reviewer as well as FAA used to write 'the officer has performed excellently, but due to number restrictions his/her ratings have been moderated'. The appraiser also used similar comment while giving average or poor ratings. In case of employees getting excellent ratings, comments were 'the officer has outstanding performance and has huge potential. He must be promoted for taking up more challenging roles'.

PMS during Subsequent Years

CTNL witnessed this cycle of performance planning and appraisal on annual basis. Issues of similar nature cropped up during subsequent years as well. Employees actual activity during the year was not clearly captured in the plan as their assignment kept on changing. Many a times, they were assigned jobs beyond the planned ones. To a larger extent performance on these unplanned jobs was also considered while rating the employees. However, appraisee had no say on this as the same was not captured anywhere. Further, performance planning would get extended up to the month of September, thereby leaving only 6 months to complete the tasks planned. Similarly, appraisal could be completed by the month of August every year. Due to this delay in performance planning and appraisal, many employees considered PMS as an annual ritual without much impact on organization.

Effort to Improve PMS

As per directive of the government, the PMS had to be based on the bell curve and therefore the same could not be scrapped out rightly by CTNL. Further, the top management at CTNL had understood limitations of the system based on their experience of its implementation during first few years (2009-2012). In order to make PMS more robust and effective, 360 degree feedback was initiated from year 2012 for mid level and senior level executives of CTNL. Every employee was supposed to be rated by 7 other employees consisting of his/her peers, subordinates; seniors across various related functions. This was done using 20 items online questionnaire.

Responses to these items (questions) were not meant to check its correctness. Rather it was intended to reflect perception of peers, subordinate and seniors about employee's performance. During first year of its implementation, respondents were submitting their feedback realistically. Majority of employees did not know the impact of their response on overall performance ratings of other employees (appraisee). However, in subsequent years employees became more aware about the system and accordingly they started responding in a socially desirable way considering favourable ratings for appraisee. This system over a period of 3 years became quite predictable and also became an annual ritual rather an effective tool of performance management and development of a healthy competitive culture within the organization.

Questions and Dilemma

Mr. Agarwal discussed the issues related to PMS with Mr. Sharma. Mr. Sharma was patiently listening to his description of the system, its accomplishments and issues, but he could not avoid certain questions that were coming into his mind such as (a) Why executives are not happy with the current system? (b) Is there any issue with its design, intent and/or implementation? (c) Should he ignore initial problems as these are normal for implementation of any new system and continue with the existing system and keep pushing for its effective implementation? (d) If certain minor changes are to be done in the system, what could be those changes? (e) Should he propose to the concerned ministry for scrapping the existing system and suggest a new one? (f) If so, what would be the features of the new system? (g) Can PMS actually help in desired cultural change?

Way Forward

Top management of CTNL was thinking to change this bell curve based PMS system and replace it with a more robust and effective system. But any change in basic principle of the system required approval by government.

The next pay revision for PSU was due from January 2017. The government constituted 3rd PRC for PSUs in June 2016. As per terms of reference, the Committee was required to review various HR issues including pay, perks, career progression etc of the public enterprises. The Committee started consulting various stakeholders. It sent letters to top performing public enterprises for their feedback on various aspects of the policy. One such letter was also received by Managing Director, CTNL. He marked this letter to Mr. Agarwal with manuscript remark '' *urgent and critical*''.

One of the important inputs asked by the Committee was related to performance management. It was an opportune time for top management of CTNL who were already contemplating to approach the government to change the existing PMS. Mr. Agarwal directed Mr. Chandra to immediately hold a 3-day workshop to brainstorm all issues related to existing PMS and future architecture of the PMS. This workshop was required to be attended by all HR Heads of the departments in CTNL. Independent consultants were also invited to attend the proceedings.

Mr. Agarwal was looking forward to receiving suggestions from these HR Heads. He was very hopeful that an innovative, efficient and practically implementable PMS system could emerge out of these brainstorming sessions which could be suggested to the 3rd PRC for incorporating in their final recommendations. He was hopeful that only such systems could help in building a healthy competitive organizational culture.

Exhibit 1: Grade wise distribution of Executives in CTNL

Grade	Number of employees
E-9	100
E-8	200
E-8	400
E-8	1000
E-7	1300
E-6	1700
E-5	2800
E-4	3500
E-3	3500

Executive in grade E-4 onward could become an Appraiser depending on the grade of the appraisee. For reviewer, minimum grade was E-6. For, FAA the minimum grade applicable was HoD in Grade E-8/E-9. For executives in Grade E-8 and above, the FAA was Director of the business vertical such as Director (Finance) for Finance vertical, Director (Operations) for Operations vertical etc.

Exhibit 2: Bell Curve and PMS

With regard to PMS, a bell curve refers to a normal distribution of performance ratings among the employees. Bell curve basically indicates shape of the distribution. Only limited numbers of employees are rated as top performers. Similarly certain percentage of employees is mandatorily rated as non-performers or poor performers. Majority of employees fall under average performer category. Due to this categorization of performance, it is also termed as forced ranking system of appraisal or relative performance measure.

Bell curve is a statistical tool based on normal distribution and therefore it must meet requirements of the distribution such as larger sample size, randomization etc.

The tool has been used to differentiate among employees performance in a more objective way. Many companies have successfully implemented the system based on this tool. However, there are companies which had difficulties in efficiently implementing it.

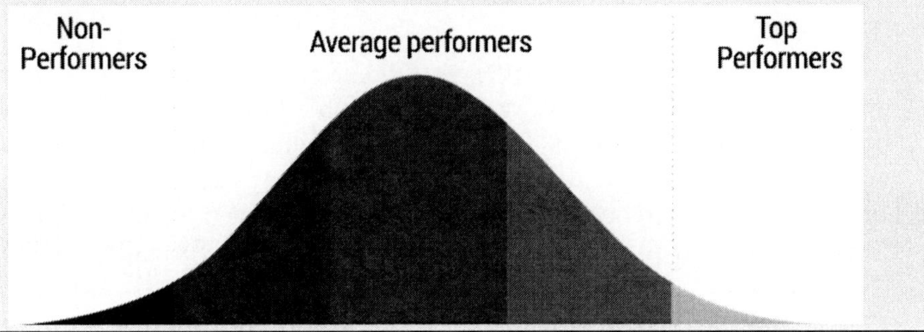

Practice Questions

1. List down the concepts of OB and HRM that can are relevant for the case?
2. How these concepts can be applied to understand various events and situations described in the case?
3. How these concepts are interrelated in the context of the case?
4. Applying these concepts, analyse various situations and events described in the case.
5. Read the case carefully and identify major issues and problems. Propose possible solution to the problem by applying the relevant concepts of OB & HRM.
6. Based on this case, develop a case let in not more than 3 pages depicting similar situation and dilemma.

Further Readings and References

Ekta, K. & Sameer, S.K. (accepted for publication), "Changing dynamics of work: Interplay of organizational and spousal support, person-job fit, work-life balance and work-family integration", ABS International Journal of Management (ISSN: 2319-684X)

Sameer, S.K. & Priyadarshi, P. (2022). Interplay of Job Characteristics, Promotion- and Prevention-Focused Job Crafting and Internal Employability. In: Anbanandam, R., Rangnekar, S. (eds) Flexibility, Innovation, and Sustainable Business. Flexible Systems Management. Springer, Singapore.

Sameer, S.K. & Priyadarshi, P. (2021). Role of Big Five personality traits in regulatory-focused job crafting. South Asian Journal of Business Studies, 10(3), 377-395, https://doi.org/10.1108/SAJBS-03-2020-0060

Sameer, S.K. & Priyadarshi, P. (2020). Interplay of organizational identification, regulatory focused job crafting and job satisfaction in management of emerging job demands: evidence from public sector enterprises. International Review of Public Administration, 26(1),73-91, https://doi.org/10.1080/12294659.2020.1848024

Sameer, S.K. & Priyadarshi, P. (2019). A Framework for managing internal employability by leveraging motivating job characteristics- Role of promotion and prevention focused job crafting. Paper presented in GLOGIFT 19-Flexibility, Innovation and Sustainable Business, an Annual International conference of the Global Institute of Flexible Systems Management (GIFT), organized by Indian Institute of Technology (IIT), Roorkee during 6-8th December 2019.

Sameer, S.K. (2022). The interplay of digitalization, organizational support, workforce agility and task performance in a blended working environment: Evidence from Indian public sector organizations. Asian Business & Management. https://doi.org/10.1057/s41291-022-00205-2

Sameer, S.K. (2022). Managing employees' performance in Indian public sector undertakings during COVID-19 pandemic-induced blended working. South Asian Journal of Business Studies, Vol. ahead-of-print No. ahead-of-print. https://doi.org/10.1108/SAJBS-05-2021-0170

Sameer, S.K. and Priyadarshi, P. (2023). "How emotionally intelligent employees manage their internal employability through role-based job crafting? – evidence from public sector enterprises, International Review of Public Administration, Vol. 28 No. 3, pp. 265-287, DOI: 10.1080/12294659.2023.2256099

Sameer, S.K. and Priyadarshi, P. (2023), "Regulatory-focused job crafting, person-job fit and internal employability–examining interrelationship and underlying mechanism", Evidence-based HRM, Vol. 11 No. 2, pp. 125-142. https://doi.org/10.1108/EBHRM-08-2021-0163

Quick Notes/ Mind Map of the Chapter

Your Ideas to add New Dimensions to the Chapter/ Case/ Anecdotes

12

The Responsible Organization

Sanjeet Kumar Sameer

Associate Professor, School of Agribusiness and Rural Management
Dr. Rajendra Prasad Central Agricultural University, Pusa, Bihar

A responsibly behaving organization creates a positive impact on society and its operating environment. Globally, there is a rising demand and expectation of people from organizations to act in manners that result into shared value and prosperity. This calls for bringing a major change in the current approach of organizations and their human resources.

Three frameworks could be popularly referred to when we intend to discuss responsible behaviour of an organization. These approaches have been highlighted considering practical significance and ability to fundamentally change the ground rules of organizational day to day working. This includes:

a) Corporate Social Responsibility (CSR)

b) Sustainable Development Goals (SDGs)

c) Environmental, Social and Governance Framework (ESG)

In this Chapter, we have provided the basic background of CSR and how it has shaped in India during past 10 years. Then, a brief understanding on SDGs has been provided covering various SDGs and Key Performance Indicators and linking these Indicators with relevant aspects of Organizational Behaviour and Human Resource Management. And finally, the concept of ESG has been explained linking how organizations need to prepare themselves for implementing in letter and spirit.

Learning from Corporate Social Responsibility Initiatives in India

Society forms an integral part of the operating environment of a business enterprise. As revenue generation by business organization is a complex process, it is highly likely that such processes may have both direct and indirect effect on the environment surrounding the components of the value chains. These effects could be detrimental in nature as well. Therefore, it is argued that such organizations should be responsible to the environment where they are operating and take actions that have a potential to generate positive externalities. Some of these initiatives such as health and education could directly help members of the society to raise their human capital, while other actions may be aimed at maintaining ecological balance. Such actions by business organizations or corporate as famously termed are refereed as Corporate Social Responsibility (CSR).

The genesis of CSR could be understood from the perspective of *Legitimacy Theory* (Shocker and Sethi, 1973). This theory is based on the assumption that there exists an implicit social contract between an organization and the society that forms its operating environment. Society has certain hopes and expectations from the organization and therefore the organization needs to take such actions that legitimize its existence. Thus, through these actions the organization would like to showcase itself as a good corporate citizen (O'Donovan, 1999).

In past, various organizations have been trying to legitimize their core activities by doing certain philanthropic works in their areas of operations. These initiatives were totally voluntary and without much monitoring and control. Globally CSR has garnered pace after financial crisis during 2008-09 (Velte, 2021). However, in India CSR activities became more than merely a philanthropic activity after enactment of Companies Act 2013 wherein a separate provision in the form of Section 135 was made in the act to regulate CSR activities. This Act converted philanthropic activity into a mandatory action to be performed by business (for profit) organizations subject to various provisions. As India aims to become a US$ 5 trillion economy in the near future and a large amount of this income is expected to be spent in the form of CSR expenditure, it is very important that usefulness of these actions are critically evaluated in a timely manner. Further, due to a change in the approach towards CSR activity from 'a choice' to 'a mandate' calls for a closer examination of its journey in India during past 8 years (since the enactment of Companies Act 2013) and analyse various associated trends. Based on these analyses, this study also attempts to propose future directions of CSR activities in India.

What does the Current Literature on CSR Indicate?

Although CSR has not been a new concept, yet there have been various definitional challenges. Organizations such as World Business Council for Sustainable Development (WBCSD) has defined CSR as "continuing commitment by business to contribute to economic development while improving the quality of life of the workforce and their families as well as of the community and society at large" (WBCSD, 1999). International Organization for Standardization (ISO) has defined it more holistically as "the responsibility of an organization for the impacts of its decisions and activities on society and the environment, resulting in ethical behaviour and transparency which contributes to sustainable development, including the health and well-being of society; takes into account the expectations of stakeholders; complies with current laws and is consistent with international standards of behaviour; and is integrated throughout the organization and implemented in its relations" (ISO, 2010). Thus, CSR invariably is aimed at improving lives of key stakeholders.

Globally, CSR has been in the forefront of corporate discussions for many years. However, it gained genuine attention post 2008-09 global financial crises. In Indian context, the philanthropic nature of CSR got converted into a 'going concern' by inclusion of section 135 in the Companies Act, 2013. The Act defined various inclusions and exclusions regarding what will be called as CSR activity, how it has to be done, what would be planning and execution process, how much fund has to be mandatorily earmarked for CSR activity, monitoring and reporting requirements etc. For example, every business enterprise that fulfils the requirements and conditions provided in Section 135 of the Companies Act, 2013 has to spend at least 2% of their average net profits made during the three previous financial years towards CSR in the current financial year. During past seven years of implementation of this Act, there is an enhanced awareness about CSR among business enterprises. Some companies have incorporated CSR plan as a part of their regular annual planning process. However, this does not hold true for all applicable organizations. The Report of the High Level Committee on Corporate Social Responsibility (HLCCSR), Ministry of Corporate Affairs, Govt. of India (2018) also provides a glimpse of mixed experiences in the India CSR domain. Various media reports are also suggesting that there is lack of compliance on CSR front, not just in terms of scope but also in terms of funding, monitoring and reporting. Issues such as sustainability of CSR intervention is grappling the academicians and practitioners related to this sector. Some of the issues related to CSR governance have been addressed in the recently published CSR. However, the actual impact of the same is yet to be seen.

Based on the extant literature on CSR, it appears that such interventions may remain relevant and sustainable in the long term if they are undertaken with long term perspective (i.e. strategic intent), are focused and consistent, are contextually aligned, poverty abhorrent and wider participation of public and private sector enterprises.

Therefore, in this context it would be worthwhile to study following key critical questions (RQs):

Q1 Has CSR intervention picked up momentum after enactment of special provisions in the Companies Act 2013?

Q2 Is CSR expenditure done for compliance or is it strategic in nature?

Q3 Is CSR expenditure focused/ Consistent?

Q4 Is CSR expenditure contextually aligned i.e. expenditure on health, education, livelihood in tune with the need/ requirement of the states i.e. is it in line with health index, literacy rate, unemployment rate?

Q5 Is CSR poverty abhorrent i.e. whether it helps in addressing the challenge of income enhancement?

a) Is CSR expenditure focused on poor states?

b) Is CSR expenditure made based on poverty considerations?

Q6 Does CSR interventions attract wider participation of public and private sector enterprises?

Exploring the Answers: Methodology

This study has been undertaken based on secondary data for the states of India on the variables mentioned in the above six critical questions. Data available from various sources including the CSR portal of the government of India has been used. State has been used as the unit of analysis in most of the cases. Answers to the questions have been explored using simple statistical tools such as correlation and regression analysis. Additionally, trend analyses have been performed to examine some of the questions namely Q2.

Results of the Analyses

Q1 Has CSR intervention picked up momentum after enactment of special provisions in the Companies Act 2013?

Before 2013, the annual total expenditure on CSR in India was in the range of Rs 3368 Cr (Bansal & Rai, 2014). However, in last 6 years i.e. 2014-2020, there has been a sharp increase in this expenditure. In fact, the average annual growth rate in CSR expenditure has been around 24%. This growth rate is much above the annual GDP growth rate of India, an indicator of economic prosperity, during the same period. Therefore, it would be prudent to suggest that CSR intervention has really picked up after the enactment of special provisions in the Companies Act 2013. The below mentioned graph explains the trend.

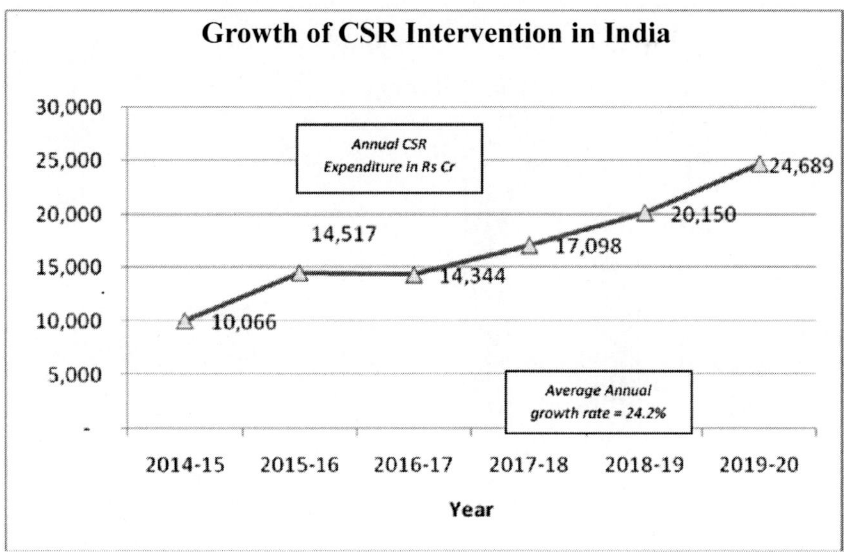

Figure 1: Growth of CSR Intervention in India

Q2 Is CSR expenditure done for compliance or is it strategic in nature?

This paper suggests a method to examine whether companies are viewing CSR expenditure as strategic or merely a compliance requirement. For example, number of companies spending on CSR activities more than that prescribed may be used as an indicator for strategic CSR while those making expenditure near to the prescribed mandate may be classified as compliance oriented CSR. Based on the aforementioned conceptualization, an attempt has been made to examine this aspect.

On an aggregate level, it appears that the CSR expenditure is strategic in nature as more number of companies are spending higher amount on CSR activities than their prescribed limits. There is no clear trend towards compliance. However, certain increase in the level of deviance from CSR mandate is also visible although this may not be so significant (r^2= 0.20).

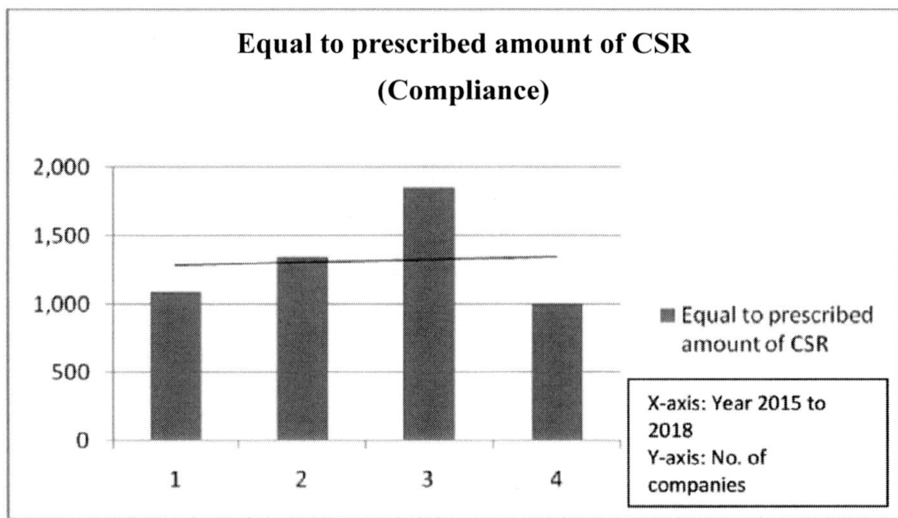

Figure 2: Compliance Orientation of CSR expenditure

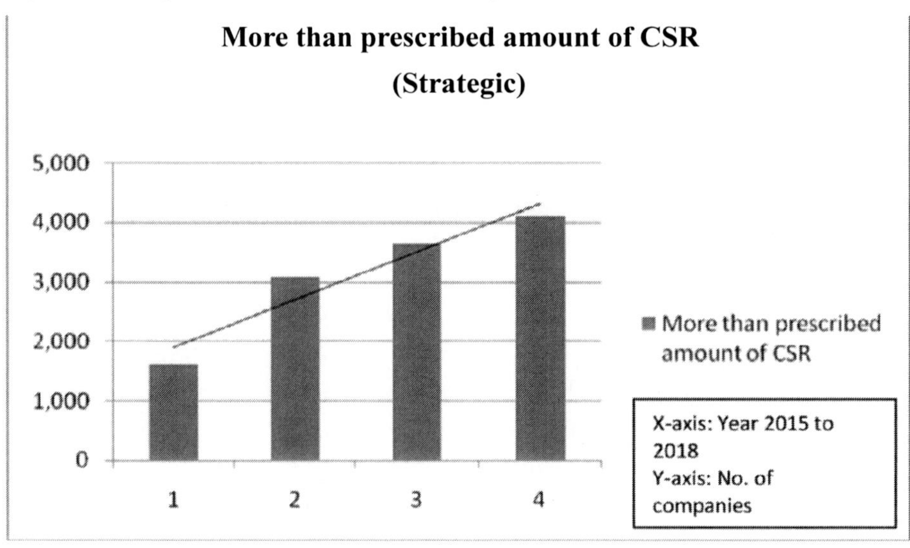

Figure 3: Strategic Orientation of CSR expenditure

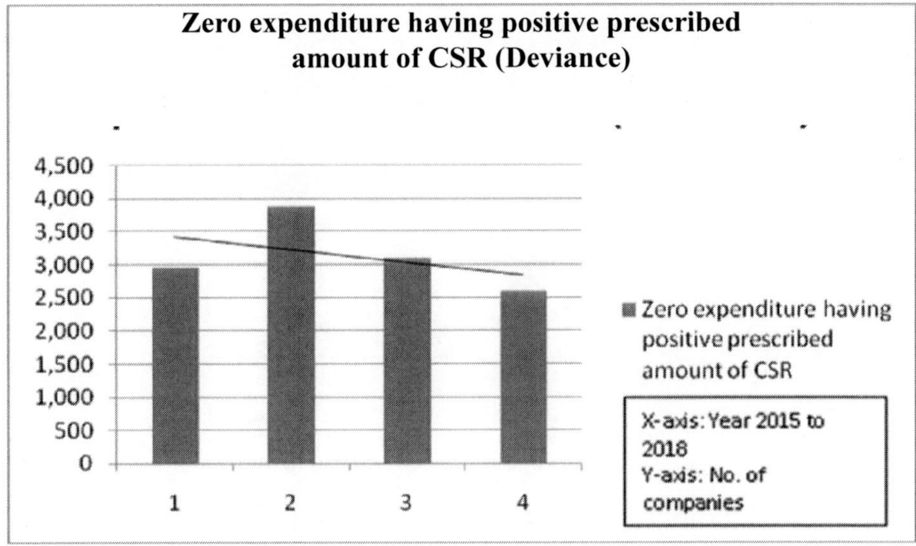

Figure 4: Deviance Orientation of CSR expenditure

Q3 Is CSR expenditure focused/ consistent?

To understand whether companies are focussed on CSR interventions, this study uses regularity and consistency in expenditure made on major CSR themes such as health care, education, rural development, livelihoods etc. over the years, say 5-7 years.

Based on the aggregated data of CSR expenditure during past 7 years, it may be concluded that bulk of expenditure (> 70%) has been made on five themes namely (a) Poverty, Eradicating Hunger, Malnutrition (5%), (b) Environmental Sustainability (6.6%), (c) Rural Development Projects (10.1%), (d) Health Care (18.8%) and (e) Education (29.5%). The consistency in expenditure on these themes has been evaluated based on trend analyses for last 7 years.

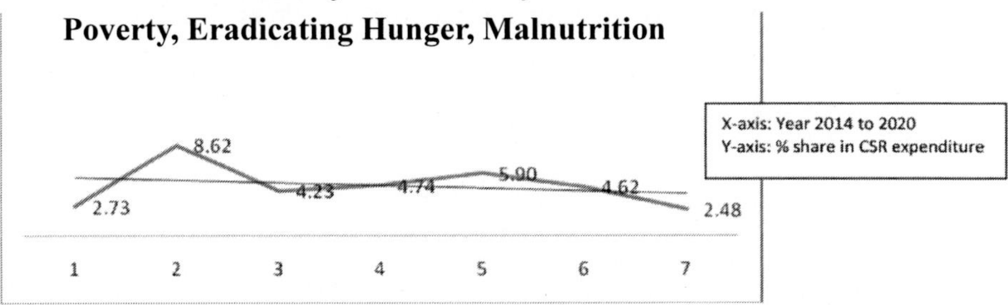

Figure 5: Percentage share of expenditure on poverty, eradicating hunger, malnutrition during 2014-2020

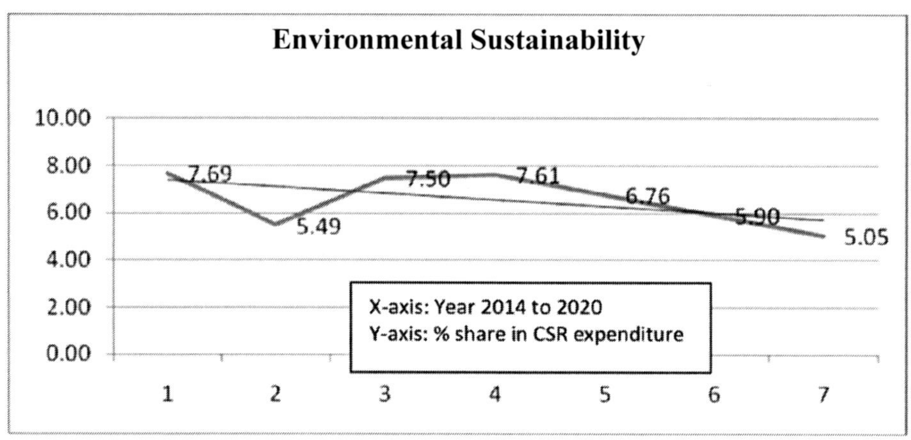

Figure 6: Percentage share of expenditure on environmental sustainability during 2014-2020

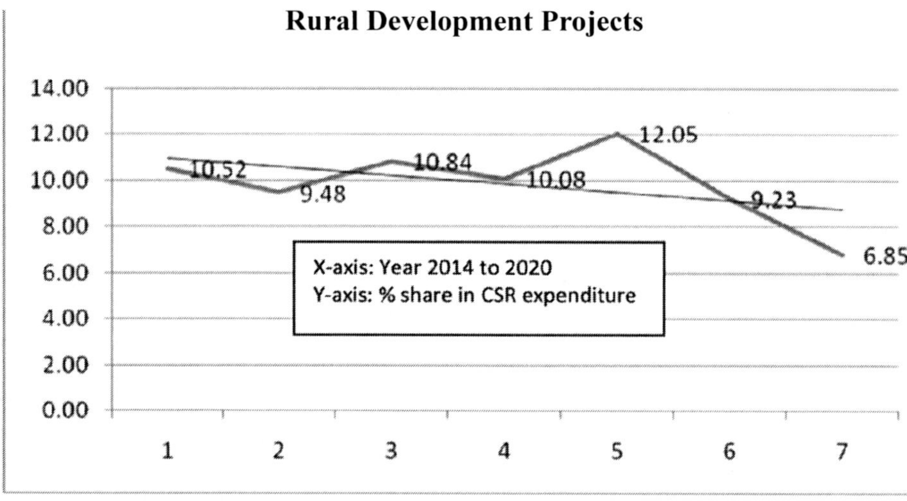

Figure 7: Percentage share of expenditure on rural development projects during 2014-2020

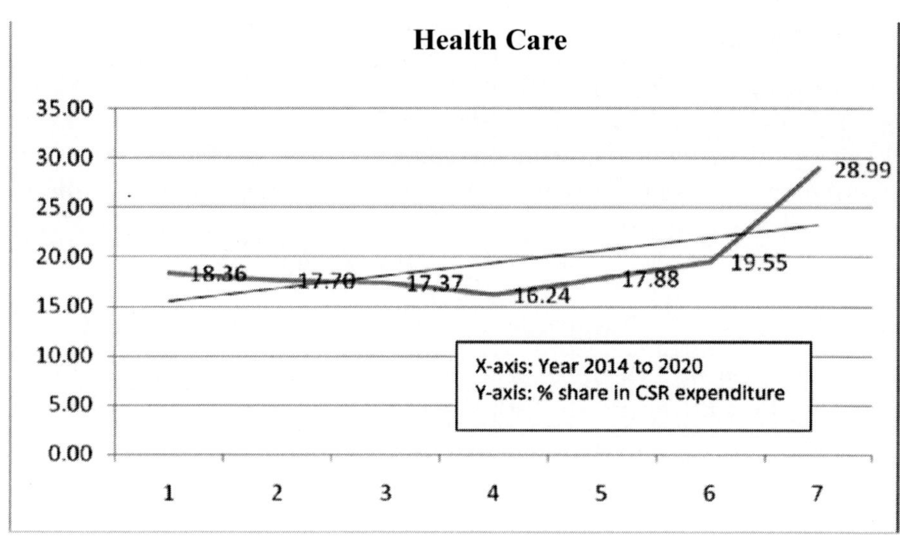

Figure 8: Percentage share of expenditure on health care projects during 2014-2020

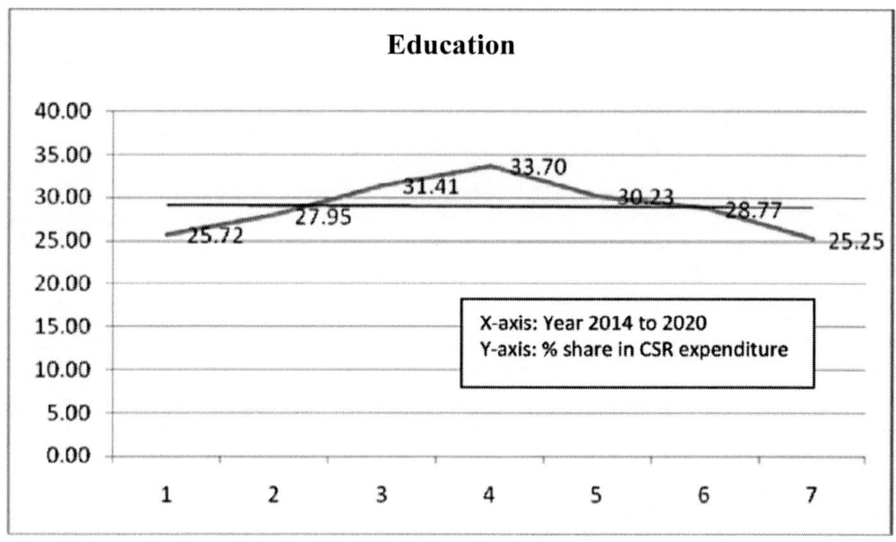

Figure 9: Percentage share of expenditure on health care projects during 2014-2020

The analyses indicate that there has been a consistent effort towards education related CSR activities. In fact, health care related CSR interventions have gained renewed focus as there is an increasing trend in expenditure on such activities. However, remaining three themes (a) Poverty, Eradicating Hunger, Malnutrition (b) Environmental Sustainability and (c) Rural Development Projects seem to be off the track.

Q3 Is CSR expenditure contextually aligned

Contextually aligned CSR intervention means that the associated activities are undertaken based on local requirements. To assess contextual alignment, data on education and livelihood promotion (through vocational training and livelihood enhancement projects) have been analysed for three states which are ranked among lowest and highest in terms of literacy rates and per capita income.

Table 1: Ranking of states based on literacy rates and per capita income (2020)

Literacy rate		**Per capita income**	
Lowest ranked states	Highest ranked states	Lowest ranked states	Highest ranked states
1.Bihar	1.Kerala	1.Bihar	1.Goa
2.Arunachal Pradesh	2.Mizoram	2.Uttar Pradesh	2.Delhi
3.Rajasthan	3.Goa	3.Manipur	3.Sikkim

Table 2: Percentage expenditure of CSR fund in poor and rich states on livelihood promotion (2020)

State	**Rank on PCI**	**% expenditure of CSR fund on livelihood promotion**
Manipur	3rd Lowest	2.25
Goa	Highest	3.22
Delhi	2nd Highest	4.44
Uttar Pradesh	2nd Lowest	5.62
Bihar	Lowest	9.55
Sikkim	3rd Highest	59.44

Table 3: Percentage expenditure of CSR fund on education in low and highly literate states (2020)

State	**Rank on literacy rate**	**% expenditure of CSR fund on education**
Bihar	Lowest	8.06
Arunachal Pradesh	2nd Lowest	9.00
Kerala	Highest	14.79
Goa	3rd Highest	38.52
Rajasthan	3rd Lowest	41.70
Mizoram	2nd Highest	82.81

From the above tables it is evident that not enough is being done on the contextually relevant issues such as poverty reduction. Similar trends are being observed in the case of two states with lowest literacy rates. However,

Rajasthan which is among three states with low literacy rates is spending bulk of CSR fund on education. The trends of six states indicate that the CSR expenditure appears to be less aligned to the contextual realities.

Q4 Is CSR poverty abhorrent?

CSR activities are aimed at resolving various social issues including poverty reduction. Therefore, it is expected that business organizations will undertake such activities in those states which require such interventions more than others. Thus, poor states are expected to get significant share in such expenditures. In other words, these interventions are expected to be poverty abhorrent. To examine this aspect, the study sought answers to following two questions:

a) Is CSR expenditure focused on poor states?

b) Is CSR expenditure made based on poverty considerations?

Gross domestic product is considered as an indicator of economic prosperity (IMF, 2020). In order to seek answer to the first question, the study used correlation analysis between CSR expenditure in the state and its gross domestic product (at current prices). This study was done on the data of past two years i.e. 2018-19 and 2019-20. The analysis suggests a strong positive correlation between CSR expenditure and state gross domestic product (correlation coefficient = 0.89). It means that companies tend to spend money where there is prosperity on gross level. To examine this aspect at individual level, further correlation analysis was done between CSR expenditure and per capita income in the state. This study shows a poor correlation (correlation coefficient = 0.10) between these two variables indicating that CSR expenditure may not have significant effect on income generation capacity of an individual (or say poverty level of individual).

Q5 Does CSR interventions attract wider participation of public and private sector enterprises.

This aspect is important considering that one of the goals of CSR is to encourage business organizations to participate in the social development process. However, it is still not been validated. Therefore, the current study examined and compared the level of participation of public sector undertakings (PSUs) and private enterprises.

Table 4: Percentage share of PSU & non-PSU in total no. of eligible enterprises for CSR expenditure

Particulars	**2014-15**	**2015-16**	**2016-17**	**2017-18**	**2018-19**	**2019-20**
No. of Non-PSU	16055	17759	19002	20980	24484	22079
No. of PSU	493	533	550	537	615	452
Grand Total (No.)	16548	18292	19552	21517	25099	22531
% Share of PSU	3.0	2.9	2.8	2.5	2.5	2.0
% Share of non-PSU	97.0	97.1	97.2	97.5	97.5	98.0

Table 5: Percentage share of CSR expenditure by PSU & non-PSU

Particulars	**2014-15**	**2015-16**	**2016-17**	**2017-18**	**2018-19**	**2019-20**
CSR by Non-PSU (Rs Cr)	7,249.11	10,302.52	11,048.37	13,447.18	15,943.97	19,447.09
CSR by PSU (Rs Cr)	2,816.82	4,214.69	3,296.03	3,650.48	4,206.30	5,241.57
Grand Total (Rs Cr)	10,065.93	14,517.21	14,344.40	17,097.66	20,150.27	24,688.66
% Share of PSU	28.0	29.0	23.0	21.4	20.9	21.2
% Share of non-PSU	72.0	71.0	77.0	78.6	79.1	78.8

It is evident from the above two Tables that PSUs although comprise only 2-3% in terms of number, yet they contribute 20.9-29% in the overall CSR expenditure. The participation level of companies could be assessed by examining the intensity of participation (IP). This paper defines IP of an enterprise as the ratio of % share of expenditure on CSR by the enterprise and its share in total number of commercial enterprises considered for CSR expenditure. This value in the case of PSU is significantly higher than that of non-PSU as evident from the following table.

Table 6: Intensity of Participation by PSU and non-PSU

Intensity of Participation	**2014-15**	**2015-16**	**2016-17**	**2017-18**	**2018-19**	**2019-20**
PSU	9.4	10.0	8.2	8.6	8.5	10.6
Non-PSU	0.74	0.73	0.79	0.81	0.81	0.80

Thus, based on higher PI it can be inferred that PSUs have better participation in CSR expenditure. One of the probable reasons could be better compliance of PSUs w.r.t. private enterprises due to overall control of the respective ministries of the government.

Broader Emerging Trends and Projections

Based on examination of various aspects of CSR journey so far in India, as aforementioned, the study identifies following key trends and issues:

a) On an aggregate level, CSR expenditure is becoming more strategic in nature than done merely for compliance purpose.

b) There is a consistency in terms of two focus areas: education and health care. However, other significant development themes such as (i) Poverty, Eradicating Hunger, Malnutrition (ii) Environmental Sustainability and(iii) Rural Development Projects seem to be off the track.

c) Enough CSR expenditure is not being made on the contextually relevant issues such as poverty reduction.

d) Bulk of CSR expenditure is concentrated in relatively prosperous states of India.

e) This study shows a poor correlation between CSR expenditure and per capita income indicating that the former may not have a significant effect on income generation capacity of an individual (or say, individual poverty level).

f) Although majority of business organizations belong to private sector, still public sector enterprises display a better participation in CSR based intervention.

g) Highly profitable and globally known enterprises, both in public and private sector, may be more concerned about CSR expenditure. However, the private player may focus more on prosperous regions of the country while public sector organizations may be undertaking CSR interventions in poor and neglected states as well.

h) Implementing CSR programmes through qualified implementing agencies (IAs) have its own advantages such as timely execution of projects provided that such IAs is efficient and are appointed in a timely manner.

i) There is an inequity in CSR expenditure across value and supply chain. Although business organizations may operate in the procurement, production, and marketing processes still they don't spend money under CSR in each component of their supply and value chain. This is one of the reasons that poor states which may be the main consumption point for the products and services generated by corporate, yet these do not figure out significantly in CSR interventions of such corporate.

Contributions to CSR literature and implications for practitioners and policy makers

The findings of the study have significant implications for academicians, practitioners, managers, and policymakers related to CSR domain. Based on legitimacy theory (Shocker and Sethi, 1973), it may be understood that various interventions under CSR made by corporate is a better way of attaining legitimacy for doing business. However, based on this study, it may be said that this legitimacy could only be attained if the interventions are socially relevant and as per the requirements of the beneficiaries. Based on findings of the study that inequity that might get perpetuated through CSR interventions by not focusing on all components of value chain, it may be posited that such interventions may further challenge the legitimacy of business enterprises. Business enterprises have an opportunity to showcase themselves as a good corporate citizen by undertaking effective CSR interventions (O'Donovan, 1999).

This study finds an encouraging trend in terms of CSR expenditure. However, this growth may have been due to lower base i.e. base effect. As mentioned in the report of HLCCSR (2018), there has been a wide variation in the prescribed and actual CSR expenditure for majority of the business enterprises. One of the possible reasons is lack of regular reporting and remedial action causing delay in project execution and completion. Therefore, CSR reporting should not be just kept for the year end but also should be part of regular quarterly reporting.

From development perspectives, particularly from the view of poverty alleviation and income generation, currently CSR intervention does not seem to be yielding an encouraging result. For example, this study finds that there has not been significant correlation between CSR expenditure on livelihood promotion programmes and per capita income. Therefore policymakers and practitioners must be careful about spending money on poorly conceptualized livelihood promotion programmes.

The study finds that there is a mismatch between contextual requirement and CSR programme implemented in a particular state. The states which are poor and resource less may like to have interventions that primarily address their acute poverty. Therefore, it is important at this stage to identify socially demanded issues based on stakeholder consultation and develop CSR programmes around those relevant issues than merely implementing projects which are easy to do, attention seeking (i.e. potential of brand visibility) and easy to showcase (i.e. of short duration). Mandating certain percentage of CSR expenditure to bottom five or ten states of the country may help in improving the coverage of the programme.

There is not much participation by private players as far as CSR interventions are and fund utilization is concerned. There is a need for a better monitoring and compliance requirement. Policymakers need to plug this loophole to further enhance their participation.

The study shows that although corporate get profit from various geographies, their CSR expenditure is primarily limited to production centres. There is a need to have more diversity in choice of regions for CSR implementation. All key actors in the supply chain and value chain should be given adequate attention. This will help in bringing equity in CSR fund utilization.

Limitations of the Study

This study aimed to examine the journey of CSR in India, identify certain key issues and suggest possible trajectories. This study examined and identified some areas of concern that might be really useful for practitioners and policymakers. However, there are certain limitations of this study. First, the current study has been undertaken considering state as a unit of analysis in most of the cases. By doing so, the local level analysis could not be performed which might have given more insightful results. One of the reasons for not doing this was lack of availability of CSR data on local level (Block / Village) for India. Second, the study did not triangulate the findings with any other primary studies. By not doing so, certain insightful findings could have been missed out. In future studies, others may use findings of this study as a supporting literature for developing a conceptual model and further testing the same through empirical studies. Due to scope limitations, only two organizations from one sector were selected for a comparative study. If similar study would have been undertaken in more number of sectors then the findings could have been more generalizable.

Conclusion

CSR is becoming more strategic in nature and therefore business planning and CSR planning shall be undertaken simultaneously. Better monitoring, implementation and governance framework could be implemented by organizations so that the money spent under CSR head has better effect. Diversity and inclusion in project scope as well as area may be given due importance to generate more impact on the society. Due consideration of various actors present in the value chain of the business should be done so that CSR expenditure does not remain limited to a few regions. Recently, Government of India has shown an eagerness to improve this sector by amending various rules related to CSR under Companies Act. It is important that these provisions are appropriately leveraged. The future trajectories of CSR may seem to be interesting.

Practice Questions

1. Based on the study mentioned in this Chapter, what do you think will be the major trend in CSR in next 5 years? Why do you think so?
2. How can organizations undertake meaningful CSR activity?
3. Can CSR be imbedded in the actions of employees of an organization? What kind of initiatves is required to be taken by an organization to do so?
4. Try to explore various sectors of the economy. How CSR can be made more strategic across these sectors?
5. Are current policy provisions of the Government sufficient for effective CSR actions?
6. What all changes do you think are required to make CSR action actually making an organization a 'Responsible Organization'?
7. Look around various forms of organizations around you. Do you think, a capable organization need to take care of other not so well organization as a part of 'CSR'?
8. Develop a case let on CSR activity by an NGO? What kinds of challenges do they face in conceptualisation and implementation?

Sustainable Development Goals, Organizations and Human Resource Management

The 2030 Agenda for Sustainable Development provides a shared blueprint for peace and prosperity for people and the planet, now and into the future. This was adopted by all United Nations Member States in 2015, A window of 15 years have been chosen to bring the defined changes globally.

A total of 17 Sustainable Development Goals (SDGs) have been identified to bring the shared global prosperity and include the following:

1. No poverty
2. Zero hunger
3. Good health and well-being
4. Quality Education
5. Gender equality
6. Clean water and sanitation
7. Affordable and clean energy
8. Decent work and economic growth
9. Industry, innovation and infrastructure

10. Reduced inequalities
11. Sustainable cities and economies
12. Responsible consumption and production
13. Climate action
14. Life below water
15. Life on land
16. Peace, justice and strong institutions
17. Partnership for the goals

To achieve these goals, a total of 169 Targets have been identified. More about Goals, Targets and Indicators associated with SDGs may be studied by accessing https://sdg.humanrights.dk/en/goals-and-targets.

What do these SDGs tell about the Role of Organizations and Human Resources?

As we minutely analyse the 17 Goals and 169 targets, we find that the role of organizations and Human resources is highly important for achieving most of these goals. For example, zero hunger can be achieved only when we have a dedicated workforce shouldering the responsibility in letter and spirit with due support of organizations. Such mammoth task cannot be achieved without active involvement of organizations. Many goals are directly related to human resources such as gender equality, good health and well being, quality education, decent work, innovation, reduced inequalities, responsible consumption, peace , justice, strong institutions, climate action etc.

What can students, early stage practitioners and organizations do to play their role in achieving SDGs?

Realising the fact that SDG is not the goal of an individual or a nation but a collective action; students in general need to see their each and every action in context of the prescription made under the SDG framework. They need to be mindful of their responsibilities and orient their action to contribute to their universal cause. Few immediate actions could be related to volunteering for awareness generation, conducting workshops and sessions among community in general, displaying mindful consumption, choosing energy efficient lifestyle etc.

From practitioner perspective, it is important to see their managerial actions somehow contributing to each of these SDGs. Thus, they are expected to formulate programmes and activities which have SDGs target inbuilt and imbedded into it. They need to assess how their products and services and corporate actions are able to contribute in fulfilling all or most of the SDGs.

From Organizational perspective, it is important to mention that the target of a nation on SDG is finally going to be achieved through organizations only. Therefore, each of the organizations needs to identify their share of responsibility and start proactively working. There is a need to have an in-house organizational level monitoring and evaluation tool to track the progress. Organizations are also required to create a climate for SDG achievement, which means that they need to generate a lot of awareness among their employees and sensitize them about their role. Incentives and disincentive structures for employees along with conducive policies and long term strategy are going to be important in organizational endeavour to contribute meaningfully to SDG goals.

Practice Questions

1. Study all 17 SDGs and list down your priority SDG. Which SDG you would personally like to contribute to and what action you intend to do as a student / early stage practitioner to achieve the goal?
2. Where do you see the role of your current organization viz. college/ university/ company/ other institution where you currently studying/ working in the process of achieving this goal?
3. Compare your SDG priorities with the priorities of your other colleagues/ peers.
4. What all components of SDGs and CSR seem interrelated? How you would like to synchronize these two frameworks?
5. Try to explore activities and things around you. List down avenues for mindful and responsible consumption? Do you think it is one of the easiest ways to contribute to the shared prosperity? Share your arguments.
6. As a practitioner, you want to take up some SDG based project but your organizational priorities do not allow you to do so. Under such condition, do you think it influences your propensity to continue within that organization?
7. Are pro-SDG organizations also pro-employee organization?

Environmental, Social and Governance (ESG) Framework

In simple term, ESG actually stands for Environmental, Social & Governance practices that an organization or an entity is expected to take care of when undertaking their key activities. Their actions should be pro-environment, community and ensure effective governance.

The term ESG is not so old. It was first used barely 20 years back in a report 'Who Cares Wins' (2004) at the joint initiative of financial institutions coordinated by the United Nations. Thus, the initial basis for ESG has been more of financial than anything else suggesting that it makes sense to invest in ESG activities for better financial performance also. As per various reported statistics, the global ESG investment has been more than US$ 30 trillion and increasing by the day. The Paris Agreement during COP21 has also expedited its pace. Europe is the torch bearer in ESG investment with almost close to 85% investment in the global ESG assets. USA is far behind with around 10% such investments. India is too trailing behind but making attempts to contribute to this space as well through sustainability oriented actions.

United Nations Environment Programme Finance Initiative and the UN Global Compact established the Principles for Responsible Investment Initiative (PRI) in 2005. In order to deeply analyse ESG issues particularly in the investment process and help organizations to invest responsibly. Globally, many countries have integrated ESG reporting as voluntary & mandatory elements. Investors/ Companies across are giving coming forward to ESG-themed/ focused investments and subsequently reporting. In India, ESG reporting is also implemented but currently is not mandated for all. This is mandatory for the top 1000 companies based on their market value on the stock exchange. This is a regulatory requirement under BRSR (Business Responsibility and Sustainability Reporting).

How is ESG measured and ESG Risk Assessed?

While measuring ESG, companies try to identify issues related to each of the three dimensions of ESG separately. Companies such as Sustainalytics have developed basic framework for assessing ESG as mentioned below (for illustration only, interested companies should contact the ESG Analysis service providers):

Category	Issue	Contribution to ESG Risk Rating
Environmental (Weightage = 43.3%)	Carbon - Own Operations	19.2%
	Resource Use	10.3%
	Emissions, Effluents and Waste	7.1%
	Environmental and Social Impact of Products and Services	6.7%
Social (Weightage= 34.1%)	Human Rights	22.8%
	Occupational Health and Safety	7.5%
	Community Relations	3.8%
Governance (Weightage= 22.6%)	Corporate Governance	11.9%
	Business Ethics	6.7%
	Human Capital	4.0%

Readers can visit https://sustainabilitymag.com/top10/top-10-esg-reporting-software to know more about ESG reporting software/ companies that do ESG analysis.

How Can we apply ESG in Organizations

Although as compliance, many organizations might be fulfilling the requirement of ESG reporting, but from long term sustenance perspective, it is important that the ESG concepts get ingrained within the organization at multiple levels. For example, it should become a part of organizational norm to look at any business opportunity from ESG angle. Organizations need to develop ESG vision and integrate it with their overall vision. Based on this, all organizational principles, processes, people and performance criteria need to be customised and rolled out.

Integrating CSR, SDG and ESG

The three frameworks of CSR, SDG and ESG basically indicate that organizations are working and performing responsibly. However, the extent of this responsible behaviour can be assessed based on the level of their activities associated with each of these frameworks. Merely doing CSR activity may not be enough. Rather one has to see to what extent these activities are adding value to the environment and society and in what manner implemented i.e. with ethical and governance principles in place. Similarly, SDGs are goal posts for the organizations and if they use the framework of ESG, it becomes easier to achieve those targets in a sustainable manner. The efforts towards integrating these concepts should be made at organizational level so that it has both required flexibility as well as framework for monitoring and evaluation and subsequently improvement.

Practice Questions

1. To what extent do you think organizations are valuing the concepts of ESG?
2. Is it really making financial sense for such organizations to implement ESG?
3. Please visit some of the organizations (including community based organizations) near your current workplace. Assess their awareness level about ESG and SDG concepts. See how they are implementing these frameworks.
4. Categorize SDGs under the three component of ESG. Where do you think there is a major challenge in implementing ESG component?
5. As a practitioner, what all initiatives you will take within your organization to implement Environmental component of ESG?
6. As a practitioner, what all initiatives you will take within your organization to implement Social component of ESG?
7. As a practitioner, what all initiatives you will take within your organization to implement Governance component of ESG?

8. Compare your ESG priorities with the priorities of your other colleagues/ peers.
9. Are pro-ESG organizations also pro-employee organization? How?

Further Readings and References

Agudelo, M. A. L., Jóhannsdóttir, L., & Davídsdóttir, B. (2019)."A literature review of the history and evolution of corporate social responsibility". International Journal of Corporate Social Responsibility, 4(1), 1-23.

Callen, T. (2020), "Gross Domestic Product: An Economy's All". Retrieved on November 20, 2021 from https://www.imf.org/external/pubs/ft/fandd/basics/gdp.htm

CSR Portal, Ministry of Corporate Affairs, Govt. of India. Data retrieved from https://www.csr.gov.in/

https://unstats.un.org/sdgs/report/2023/The-Sustainable-Development-Goals-Report-2023.pdf

https://www.deloitte.com/global/en/Industries/mining-metals/perspectives/embedding-esg-into-organizations.html

https://www.sustainalytics.com/corporate-solutions/esg-solutions/top-rated-companies

Indian States by GDP (n.d.), Data retrieved fromhttps://statisticstimes.com/economy/india/indian-states-gdp.php

Islam, M. A. (2017). "CSR reporting and legitimacy theory: Some thoughts on future research agenda". In The Dynamics of Corporate Social Responsibility (323-339).Springer, Cham.

Ministry of Corporate Affairs, Govt. of India (2018), Report of the High Level Committee on Corporate Social Responsibility, Report accessed on November 20, 2021 from https://www.mca.gov.in/Ministry/pdf/CSRHLC_13092019.pdf

Rai, S., & Bansal, S. (2014). "An analysis of corporate social responsibility expenditure in India".Economic and Political Weekly, 49 (50). Retrieved from http://www.epw.in/node/130387/pdf

Shocker, A. D., and Sethi, S. P. (1974), "An approach to incorporating social preferences in developing corporate action strategies".inSethi, S. P. (Ed)., The unstable ground: corporate social policy in a dynamic society, 67-80, Melville, Ac, London.

Sinha, S.N. (2021), "Corporate Social Responsibility in India- A Case of Government Overregulation?", Economic & Political Weekly, 56 (26-27)

Velte, P. (2021). "Meta-analyses on corporate social responsibility (CSR): A literature review". Management Review Quarterly, 1-49.

World Business Council for Sustainable Development(1999),CSR definition, Retrieved from http://www.wbcsd.org/work-program/business-role/previous-work/corporate-social-responsibility.aspx.

Quick Notes/ Mind Map of the Chapter

Your Ideas to add New Dimensions to the Chapter/ Case/ Anecdotes

Quick Notes/ Mind Map of the Chapter